Culture and Climate Change: Recordings

Edited by Robert Butler, Eleanor Margolies,
Joe Smith and Renata Tyszczuk

Published by Shed, Cambridge

Editors: Robert Butler, Eleanor Margolies,
Joe Smith and Renata Tyszczuk
Managing editor: Vicky Long
Design by Hyperkit

ISBN 978-0-9557534-1-1

Printed by Lecturis, Eindhoven

This publication is printed on Munken Lynx paper which is FSC accredited and produced at a mill certified to the ISO 14001 and EMAS environmental management standards. Printed at Lecturis using eco-friendly BIO inks.

FSC www.fsc.org MIX Paper from responsible sources FSC® C008041

The Open University

THE ASHDEN TRUST

Contents

Acknowledgements

We gratefully acknowledge the Ashden Trust and the Open University OpenSpace Research Centre for their support for the discussions and this publication.

The discussions were recorded in collaboration with TippingPoint, the University of Oxford, the Eden Project, the Open University and the National Theatre Studio. Our thanks are due to each of these organisations for their encouragement and support and for providing inspirational venues, and to the audiences at the four recordings.

Catherine Chambers, Matthew Holley, John Sinton and Kristoff van Leeuwen at the Open University managed the recording, editing and post-production of the podcast.

Wallace Heim, Benjamin Morris, Kellie Payne and Bradon Smith helped us to map out the territory of the discussions and contributed to the annotations and timeline. Peter Gingold and S. I. Martin kindly read and commented on the book in draft form.

We would particularly like to thank all the contributors to the discussions, who shared their knowledge and expertise so willingly.

Robert, Joe and Renata

Introduction

It's a subject that touches on some of the great themes of art: love, empathy, generosity, sorrow, greed, waste, loss, suffering and want. It raises significant ethical and moral questions in terms of how humans think about their place in the world. It concerns the great philosophical questions of time, justice, custom and dwelling. And it has the potential to affect every aspect of life – from the food we eat to the homes we build. The discovery that the way we are living is endangering our future and the futures of those who follow us might be expected to resonate through all forms of cultural expression.

Indeed, in recent years, an increasing number of exhibitions, performances and publications have presented cultural responses to climate change. Many of these are listed in the timeline at the back. But is this really something new? Or are we simply reinterpreting long-established themes – humanity and nature, apocalypse and utopia, hubris and the gods – that reach back beyond the Ancient Greeks?

Over the last decade, three of us – a theatre critic, a geographer and an architect – have followed this field of work and the critical response. It has been fascinating to watch the subject develop, but we have also shared feelings of frustration. We've noted some confusion about how to 'place' the work and about how it relates to science, policy, and the creative arts more generally. We've sometimes found ourselves taking part in unproductive debates about whether particular cultural responses to climate change should be defined as communication, activism or artistic experiment. And yet, looking beyond labels, we were aware that there was already a rich history of cultural response to changing physical environments, and that contemporary work might connect to that tradition in many different ways.

It struck us that there was a need for greater clarity about how these activities are discussed, reviewed and, even, funded. We felt, in short, that 'culture and climate change' needed mapping. We wanted to sketch out a critical framework that would include the context of the work, where it comes from, what makes it distinctive and where it might go. So we devised a series of discussions on the subject, inviting artists and academics from many disciplines, as well as producers and journalists. We knew from the calibre of those taking part that they would bring expert observations to a discussion that was still finding its way.

We decided on four topics: first, the history of cultural responses to climate change; second, popular culture and mass media; third, the anatomy of works in this area; and fourth, the way that culture, politics and science interact as we imagine and respond to scenarios of the future.

The discussions were recorded in autumn 2010 at a TippingPoint conference in Oxford, at the Eden Project in Cornwall, at the Open University's offices in London, and at the National Theatre Studio in London. They are available as a podcast under the heading of 'Culture and Climate Change' on the Open University's iTunes U platform, and represent a first extended audio exploration of this subject to be made publicly available. Over a six-month period, from the beginning of December 2010, over 10,600 tracks have been downloaded.

As we listened to the original discussions, we became aware that the speakers were referring to a great hinterland of cultural and scientific work which they didn't necessarily name or describe. We decided to produce a book which would make that background context more visible and worked with an editor to annotate the conversations and to provide a timeline and list of suggested reading and viewing – amplifying the points raised, pinning down references and suggesting further avenues of enquiry. Full transcripts are available from iTunes U; here, they have been condensed to about half their original length. The hesitations and repetitions of ordinary speech have been edited out, and the speakers have, in one or two places, added a few words to clarify their original meaning.

This is work in progress. The timeline and notes focus on British and North American work, reflecting our own experience and networks. The publication does not aim to be definitive. Nevertheless, it represents a range of people who are thinking hard about this subject. We hope others will respond to the podcast, and this publication, and take the conversation forward.

To download the podcast, visit http://open.edu/itunes/subjects/environment-development-and-international-studies.

To download a PDF of this publication, and to follow future developments, visit www.ashdenizen.blogspot.com.

Three essays

Looking for the tune

Robert Butler

One evening when I was 15 or 16, I took a book down from my parents' shelves which had an appealing title for a teenager: *The Romantic Rebellion*. It turned out that this book by the art historian Kenneth Clark dealt with the battle between romantic and classic art in the years before and after the French Revolution. Within the first three pages, Clark describes Jacques-Louis David's painting *Oath of the Horatii* in which the three sons of Horatio, the Horatii, take an oath on their swords in front of their father. From the black and white illustration, it looked a very formal old-fashioned picture, but Clark said that when this was painted in 1784, five years before the French Revolution, it was 'a perfectly conscious piece of propaganda, which was planned like a political campaign'. I was hooked.

Till then, I had mainly thought of great works of art as representing the values of princes and patrons. But Clark described the paintings of this period in a way that was new to me: they were part of the turbulence of the age. He compared this severe classical-looking painting with Picasso's large symbolic mural *Guernica*, painted after the aerial bombing of the Spanish town by the German Luftwaffe. 'Suddenly, through a work of art,' he wrote, 'men became conscious of moral responsibility.' Two pages later, Clark described *The Death of Marat*, where the radical French journalist lies murdered in his bathtub, as the 'greatest political picture ever painted'. He said that the impact of the painting lay in David's powers of 'concentration' and 'elimination', but the aspect of Clark's book that kept me turning the pages was the suggestion that the revolutionary ideas of an age were reflected by, and — in part — shaped by, the leading artists of the day.

It was 15 years after reading *The Romantic Rebellion* that I, like many others, became aware of a scientific discovery that would precipitate another shift in human values: anthropogenic climate change. The ringing sounds of *liberté, égalité* and *fraternité* were replaced by the more prosaic-sounding 'greenhouse gases', 'CO_2 emissions' and 'carbon footprints', but there was a political dimension that the two had in common: at the centre of the debate over climate change lay the balance of power between rich and poor. The reports from the Intergovernmental Panel on Climate Climate (IPCC), which had been released with an increasing sense of urgency in 1990, 1995, 2001 and 2007, warned of developments that might be as convulsive and far-reaching as the French Revolution.

Single events can alter our world view. When the United States dropped an atom bomb on Hiroshima, a Japanese city of 250,000 people, at 8.15am on 6 August 1945, the news affected the way people in every country thought about the world and their futures in it. The impact of that explosion can be traced in thousands of ways in novels, paintings, plays and ballets. So too, in their own way, the release of the IPCC reports has raised new questions about how we relate to one another (how countries and continents might relate, how generations might relate), what our responsibilities might be, and what might happen if concerted action is not taken. The stakes for our future look about as high as they could be. In this new age of turbulence, how will artists address issues that demand a new sense of moral responsibility? Who will create the next *Oath of the Horatii*?

This question applies to all forms of art. There have been many moments in Western drama, for instance, when what has happened on stage has resonated powerfully with what is happening in the world outside. Aeschylus's play *The Persians* (472BC) is the only Greek tragedy (of the 32 that survive), to depict a historical event, the defeat of Xerxes' forces by the Greeks, eight years earlier. Part of the playwright's genius was to show the event from the Persians' point of view — a shock, no doubt, for many in the Greek audience.

In the nineteenth century, the role of women in society was given intense scrutiny in Ibsen's plays. When Nora walked out of her home, at the end of *A Doll's House* (1879), leaving behind her husband and children, the slam of the door was said to have reverberated across Europe. In more recent times,

John Osborne's *Look Back in Anger* (1956) provided a focal point for truculent antipathy towards post-war conservatism. More recently still, there have been powerful plays on civil rights, AIDS, and the wars in Iraq and Afghanistan. But there have been very few plays on environmental themes. This is remarkable as environmentalism can be said to be the most significant new strand of political thought since the Second World War. The widespread understanding that the world's resources are finite, and that we are endangering a delicately balanced ecosystem, owes much to Rachel Carson's account of pesticides in the food-chain in *Silent Spring* (1962), to the iconic photo of the earth taken from space (1972), to the oil crises, and the discovery of the hole in the ozone layer (1985). These moments, along with many others, have changed our sense of where things come from and where they go. Now people talk, in a way that would have amazed people 20 years ago, about water as 'the new oil' and waste as 'the new gold'.

This shift in perspective was given its most decisive push by the publication of the IPCC reports and by Al Gore's film *An Inconvenient Truth*, which popularised the findings of those reports. For many people, the news about climate change has altered their outlook as decisively as Darwin's *Origin of Species* did for the Victorians. But there is a difference: Darwin's discoveries fitted in with the dominant narrative of industrial progress where those who worked hardest flourished best. The implications of climate change run up against the very powerful emotions that cluster around individualism, free markets, and rising personal consumption. It may be one reason why the artistic response has been so much slower.

Privately, theatre directors have given me many reasons for not staging plays that explore this shift in perspective. 'The stories are no good.' 'There's no ending.' 'It's not immediate.' 'I don't know anyone it has affected.' 'I don't want to address an issue.' 'I don't want to educate.' 'It's middle-class — it's not for the poor.' 'People think it's another way to tell them what to do.' 'Is it happening? Is it true?' 'We grew up with the fear of nuclear war and that didn't happen.' And so on. The relationship that theatre has, or doesn't have, with this subject is indicative of a much wider dissonance. In looking at culture and climate change in the essays and conversations gathered here we are not merely considering works that are 'about' climate change, we are also considering how the idea of climate change indirectly informs the culture of our time — how the idea is resisted as well as embraced — just as Darwin's ideas informed, for instance, the novels of George Eliot and Thomas Hardy. So far, we have only seen the first efforts by artists and writers to come to terms with the full meaning of the climate science that has emerged over the last 20 years. These are still early days.

Sixteen years before the Russian Revolution, Maxim Gorky wrote a play that prefigured the events in his country which would change twentieth-century history. In *Philistines* (1901) Gorky depicts a bourgeois family that ferments with rage under the rule of a narrow-minded authoritarian father. When the original production received its dress rehearsal in St Petersburg in 1902, the play's naturalism was considered so provocative that a detachment of police was ordered to the area and the cavalry was drawn up in front of the theatre. In one of the play's most telling speeches, Teterev, the intellectual alcoholic, catches the whole idea of living through the early stages of a turbulent period, when profound changes are underway and no-one can quite guess what is about to happen. Teterev compares it to the moment when musicians warm up before a concert: 'You can hear fragments, distinct little pieces. A trumpet suddenly. Violins ... every now and then a lovely phrase. And in all that disorder you begin to look for the tune. What will they play when they settle into it?'

Why climate change is different: six elements that are shaping the new cultural politics

Joe Smith

In one of many acute moments in Ian McEwan's novel *Solar*, the Nobel physicist Michael Beard finds himself at the receiving end of a lecture on climate change from a fellow scientist: 'Beard sank into a gloom of inattention,' McEwan writes, 'not because the *planet* was in peril — that moronic word again — but because someone was telling him it was with such enthusiasm' (McEwan 2010: 36).

If we want to understand why we continue to behave in ways that contradict good quality scientific knowledge, we must consider exactly what is novel about the cultural politics of climate change. By 'cultural politics' I mean the ways in which the values and meanings that underpin our economics, politics and society are generated and argued over. Much current discussion about climate change falls between the overstated rhetoric of jeopardy, which is now having a diminishing impact, and more sober discussions round risk and uncertainty, which are largely unreported. Here I want to draw out some features of the cultural context that surrounds media reporting and artistic work on climate change.

Climate change has produced many unexpected responses, one of which (as Renata Tyszczuk points out — see page 25) resembles Stockholm syndrome — the phenomenon of hostages becoming emotionally attached to their captors. A good deal of discussion about climate science and policy has an excited, even breathless tone as it conjures images of social and ecological jeopardy, wrapped up in sober scientific prediction. NGOs and commentators argue that devastation is inevitable unless action is taken in response to specific figures. For example, the website for the network 350.org suggests that '350 is the most important number in the world—it's what scientists say is the safe upper limit for carbon dioxide in the atmosphere... the planet face[s] both human and natural disaster if atmospheric concentrations of CO_2 [remain] above 350 parts per million'. Andrew Simms, who writes a monthly blog for the One Hundred Months campaign, argues that time is 'fast running out to stop irreversible climate change... We have only 100 months to avoid disaster'.

Insistent arguments such as these have been allied to a very simplified representation of the state of climate science. Phrases such as 'the science is finished' or references to 'the IPCC consensus' have been used to foreclose debate, so that everyone has to move on to the next stage — taking action. For example, as Oliver Morton points out in the Futures discussion (page 86), the notion of a 'tipping point' often functions as a rhetorical trump card. Novelist Giles Foden — whose latest novel, *Turbulence*, deals with the special significance of meteorology for the 1944 D-Day landings — suggests that there is 'a kind of hubris' in the reference to 'tipping points': 'it invests too much in human predictions of the nature and consequences and scope of the event'. He suggests that the doom-laden term might be replaced by other metaphors 'which are generative and work positively as an invitation to action' (from an unpublished paper). Similarly, research suggests that taking short cuts to public attention through disaster imagery — such as photos of drowning polar bears or drought-stricken children — delivers diminishing returns in terms of political engagement, as well as carrying other kinds of costs in terms of the dignity of the subject and our relationship to it (see, for example, Cohen, 2000).

Over recent years, there have been signs of a falling away of public engagement even as the climate science community become more confident of its results (see, for example, Spence et al, 2010). This is in part because the phrase 'climate change' is put to work in complex ways, and the issue generates multi-layered cultural politics. This was demonstrated — perhaps unwittingly — by the Climate Camp protestors objecting to a proposed third runway at Heathrow airport who held up large scale portraits of potential climate victims from around the world alongside a large banner stating, 'We Are Armed Only With Peer Reviewed Science'. In his pioneering examination of these issues in *Why We Disagree About Climate Change*, Mike

Hulme suggests that climate change has become 'an idea that now travels well beyond its origins in the natural sciences. And as this idea meets new cultures on its travels... [it] takes on new meanings and serves new purposes' (2009, xxvi).

Great progress has been made in communicating the key points about the science and politics of climate change — some of these points are marked on the timeline found at the back of this book. In just two decades, awareness of the basic contours of the issue has spread across the globe. It is now widely accepted that human activity is almost certainly altering the climate, and there is also recognition of some of the key physical hazards, including melting of polar ice sheets, sea level rise, and the potential for drought, storms and floods around the world. There is also increasing recognition of climate-related threats to economic and social stability in terms of food, energy, water and quality of life. In other words, when it comes to humanity's responses to climate change there is a shared sense across the world that the stakes are high.

However, it is also true that western societies remain locked in a dogged commitment to carbon-based economies and lifestyles, displaying 'a gambler-like tendency to commit to failing bids, and to continue with small and incremental adjustments even in the face of group calamity' (O'Riordan *et al.*, 2011: 7).

So what is novel about the cultural politics of climate change? The novelty lies perhaps not in any one of the following six features but in their combination.

The first distinguishing feature is global pervasiveness: climate change gets everywhere — from doorsteps to boardrooms — and pervades all layers of formal politics from parish and local councils to parliaments and international conference halls. It reaches across the world and across generations in ways that no other public policy concern does. This dimension of the issue is frequently noted in both popular and professional contexts, but not the distinctive challenges it poses in terms of ethical and political debate or cultural responses.

A second element is uncertainty, in both science and policy. Media representations in the past have more often than not failed to acknowledge that the sciences of global environmental change are not just 'unfinished' but 'unfinishable'. Climate change research is not unique in this respect, but it is a particularly dramatic and important example of what Funtowicz and Ravetz have termed 'post-normal science'

(2003). Climate change should not be responded to as a body of 'facts' to be acted upon (with the IPCC as prime arbiter), but might instead be considered as a substantial and urgent collective risk management problem. Projecting climate change as a risk problem rather than a communication-of-fact problem helpfully deflates 'debates' about whether climate change is or isn't a scientific fact. Such an approach doesn't walk away from the science but rather opens more possibilities for people to be tolerant of the unsettled, developing relations between climate science, policy and politics.

Thirdly, knowledge of climate change emphasises the interdependencies between human and non-human systems, both near and far. Acknowledgement of humanity's state of interdependence can be traced back at least as far as the depiction of city life as dependent on its rural hinterland in Virgil's *Eclogues*, written over two thousand years ago. There have been numerous invocations of interdependence across the last century in relation to, for example, food and farming, civil rights and biodiversity. However, climate change calls up interdependence both as a description of environmental processes (e.g. relating to the consequences of the release of anthropogenic greenhouse gases) and, inextricably, as a political problem (see Smith *et al.*, 2007).

The potential for substantial changes in earth systems that we have tended in the past to think of as stable or static, forces us to acknowledge that we live on a dynamic earth. It would be a mistake, nevertheless, to replace the hubristic assumption of human separateness from nature with an account of evenly balanced interdependence between the natural and the human. Acknowledging our new place in the world includes understanding and respecting the difference between truly interdependent relations and those 'earthly imperatives' which might have huge consequence for humans, but not for nature. A cultural politics that is rooted in a rich understanding of global environmental change is likely to look quite different from our current state. As Nigel Clark puts it in his book *Inhuman Nature*, we are 'still a long way from the cosmopolitan thought we need, the kind that might point the way to forms of justice and hospitality fitting for a planet that rips away its support from time to time' (Clark, 2010: 219).

It is also important to note that interdependency does not imply an uncomplicated convergence of interests around action. This leads to my fourth point: if the cultural politics of climate change echoes post-colonial discourse, by paying attention to history and notions of 'vulnerability' and 'responsibility', it is with good reason. The fossil-fuelled development of the last century shaped individual life chances and national opportunities for good and ill across the planet, but these chances were patterned by the pre-existing political economy of development. When Arctic Inuit assert their 'right to be cold', and Pacific Islanders argue for action to protect their land from rising sea levels, they do so in the knowledge that the threats they face have been generated by the rich world's exploitation and consumption of resources over centuries. These questions about ethics of responsibility and vulnerability serve to shift the boundaries of political community. However, there is a danger of complacency in the assumption that climate change means 'there is no other way', and that we will inevitably 'form a global community with a set of shared beliefs', as Tim Flannery suggested in a recent interview (Flannery, 2011). It seems likely that international climate politics will become far more antagonistic in the future. This need not halt progress on climate change action: rather, it may help to generate the 'real', honest and urgent politics that has long been lacking.

The fifth distinctive feature is the interdisciplinary nature of the knowledge upon which climate change science is founded. As one climate expert remarked in 1961, 'The fact that there are so many disciplines involved, as for instance meteorology, oceanography, geography, hydrology, geology and glaciology, plant ecology and vegetation history — to mention only some — has made it impossible to work... with common and well established definitions and methods' (quoted in Weart, 2008: 33). The processes of the IPCC represent one of the most ambitious attempts at global peer review of a specific set of questions, and draw together a very broad body of scientific research. The panel's reports summarise an extraordinary body of intellectual achievement. However, even that process is limited by its failure to integrate the social sciences, arts and humanities. This is all the more surprising given how heavily the processes of the IPCC, as well as of the United Nations Framework Convention on Climate Change, rely on 'scenarios', and hence involve acts of imagination about possible futures in human as well as natural systems. One task for cultural work in this area is to open up thinking about what it means to construct imagined futures, and the intellectual and creative work it might require. This demands a new relation with time. Just as climate change prompts us to extend the boundaries of politics in space, it also requires that we extend it in time.

The distinctive temporalities invoked by thinking and talking about climate change represent the sixth distinctive feature of the cultural politics of climate change. Economists and policy specialists have sought ways to give future generations a voice in the present, albeit through very attenuated or clumsy proxies such as 'discount rates' and 'policy targets'. Past generations can also be heard: from our prehistoric ancestors, who coped with earlier changes in climate with doggedness and imagination, to the more recent ancestors who bequeathed inventions and discoveries that have changed both climate and our understanding of it, such as steam engines or techniques for retrieving and interpreting ice cores. Although contemporary human interests are more audible than those of the past, this expanded ethical, political and cultural community is increasingly present in our thoughts and actions. As Mike Hulme says in the discussion which is recorded here, the future 'is a place that we all live in, in our imaginations'.

Mike Hulme goes on to point out that climate change is 'both political and cultural' (page 76). It is the meeting points between these spheres which we have sought to explore in the four discussions documented in this publication. The six features I have outlined here — global pervasiveness, uncertainty, interdependency, the reverberations of history, interdisciplinarity and temporality — are shaping a new cultural politics which is, like climate itself, in a permanent state of change.

On constructing for the unforeseen

Renata Tyszczuk

When confronted by global environmental change, humanity has tended to try to forestall the predicted events — to stop the future happening. We tend to speak as if the solid certainties of our civilisation are under threat — a way of thinking that produces little more than anxiety. An alternative response might be to think in terms of 'improvisation': discussions around climate change might then serve as a context for exploring potential future worlds, opening up more promising possibilities for living on a fragile — for humans — and dynamic earth.

Our situation is precarious; several decades of earth systems science confirm that we live on an unpredictable planet. As Nigel Clark reminds us, the earth is astonishingly good at unleashing primordial forces of destruction and pulling the ground from under our feet (Clark, 2010). In the 'History' conversation documented in this book, he suggests that if we look back in time, we see evidence of our ancestors' resourceful responses to abrupt changes in climate and other natural disasters. Human existence on the planet has always been provisional; instability is in the nature of our existence.

Many of the systems and categories we assume to be fixed — whether economic, social, ecological or political — are actually 'conditions' or 'constructions', which are changeable. In recent years, the concepts of progress, modernity, the quest for certainty and the supposedly unshakeable foundations of knowledge have been questioned by philosophers such as Stephen Toulmin, who describes 'the scaffolding of modernity' as 'a set of provisional and speculative half-truths' (Toulmin 1992: 116).

As all human constructions are inherently fragile, they need to be approached with caution and an acknowledgement of responsibility. They need our continued care and attention. This care extends to the monsters we have created as much as to the monuments to human ingenuity. After all, in Mary Shelley's story, Frankenstein's creature is made furious by neglect — and only then becomes a danger. Furthermore, the contemporary acknowledgement of the intersection of environmental change with economic, technological and cultural globalization is leading to changes that are especially unpredictable (see Smith *et al.*, 2007). How can we prepare ourselves, both practically and imaginatively, for the unforeseen?

In a note, the philosopher Ludwig Wittgenstein describes the construction of the idea of progress:

> Our civilization is characterized by the word 'progress.' 'Progress' is its form rather than making progress being one of its features. *Typically it constructs*. It is occupied with building an ever more complicated structure... I am not interested in constructing a building, so much as in having a perspicuous view of the foundations of possible buildings. (Wittgenstein, 1980: 7; my emphasis)

Wittgenstein suggests that there is a way of thinking otherwise for all of our constructions. What 'might be' if our social, physical and intellectual circumstances were different? How can we look at the 'foundations of possible buildings'?

Accepting the possibility that our constructions might be provisional not only allows us to 'inhabit' instability, but also to respond imaginatively to unexpected developments. The word 'improvise' comes into English from the Latin, *improvisus*, 'unforeseen', via French and Italian phrases describing the spontaneous composition of verse or music. The discussions gathered in this book touch on the importance of approaching our current situation with a sense of adaptability or improvisation. In the 'Futures' discussion, for instance, Carolyn Steel talks of the need for mental 'openness' rather than preparedness, and suggests that this quality might be fostered by art. In the 'History' discussion, Wallace Heim points to the intellectual history of the idea of improvisation, referring to Aristotle's notion of 'practical wisdom' or *phronesis* — which characterises the ability to reflect on and choose a mode of action in unforeseen situations.

We are a generation which seems to understand its place in time and in history as defined by climate change. This has provoked a range of confusing and contradictory responses. Most writing and images around mitigating and adapting to climate change fall into the categories of policy-world propriety or aggressive campaigning rhetoric. The result is an odd mix of bureaucracy and terror. Our collective imaginations are poised at the edge of a precipice no one particularly wants to look over. Little wonder that the political and social response to climate change is so mixed: disengagement seems an obvious response. Environmental issues are viewed as intractable, regardless of what we do, and at the same time we have become maybe a little bit too comfortable with the bad news stories that keep on coming. We think of ourselves as constantly at risk of being overwhelmed by forces beyond our control. Sometimes we behave as if the anthropogenic components of global environmental change have taken us completely by surprise. We are adept at turning a blind eye to the blindingly obvious. This stasis around climate change discourse has carried on for at least two decades and we have become used to the feeling of being stuck. It is as if we are unable to make our own choices in spite of good information that offers routes to self- preservation. In short, we are held captive by our own fears and misgivings and yet grateful for the small mercy of continued survival: a climate change equivalent of Stockholm syndrome. Like the hostages in the 1973 bank robbery, we have started to show affection for the thing that is trapping us. Perhaps our actions in the present moment need to be more responsive to the potential of our precarious human world. This may entail taking steps that are more 'provisional', less certain and fixed.

The rumblings of revolutions and earthquakes in 2011 give us plenty to think about. Earthquakes, tsunamis, nuclear disasters and tyrant-toppling, with all their human-displacing, species-endangering, world-ending implications, awaken apocalyptic imaginings and give rise to unspoken anxieties. Both the most sudden political hurricanes and the slowest biogeochemical cycles can remind us that the present is uncertain, contingent.

Political revolutions are an interim state; the only certainty is that life will be different and we will adapt in many different ways. As Rebecca Solnit writes, introducing a discussion of the unfinished revolutions in Tunisia, Egypt, and Libya: 'Revolution is as unpredictable as an earthquake and as beautiful as spring. Its coming is always a surprise but its nature should not be' (Solnit, 2011). Meanwhile, the 'once in a thousand years' quake in Japan leaves behind long-lasting damage, areas contaminated and uninhabitable for years and reactors entombed for all future history.

Unforeseen disasters defy containment by any system set in place prior to their occurrence. In the face of such disasters, as Michel Serres puts it, 'a thousand useless ties come undone' (Serres, 1995: 124). Society thinks about itself differently through such shocks and aftershocks. Can we therefore begin to adopt an improvisational agency by confronting a human temporality that is about 'living in a permanent earthquake' and stretches back into forgotten pasts and on into unforeseen futures?

Provisional solutions can be temporary, unfinished, or just 'for the time-being'. But they can also be anticipatory of the 'not yet' or 'still to come'. The provisional can be thought of as a draft, a sketch or a sample of something new, an interim procedure or a process of trial and error. Above all, it implies questioning, straying off existing paths and finding others, and a tolerance of failure. The sculptor Richard Wentworth writes:

> I grew up in a world held together with string and brown paper and sealing wax, and that's how it was. I slowly realized that this is the *underlying condition of the world*, and there's nothing I like more than when, for example, there's been a near-disaster at NASA and they say: 'If it hadn't been for the chewing gum...'. It's not because I want to fetishise chewing gum or the aesthetics of gum pressed over some break or membrane; it's because *we have the intelligence* to think: 'hey, there's a malleable, mastic material and we can use that.' A large part of our lives is spent using that very edgy bit of our intelligence. (quoted in Henry, 2007; my emphasis)

Wentworth's reference to NASA recalls one of the best-known examples of human ingenuity and improvisation (supported by technological prowess and rehearsal). On the Apollo 13 mission in 1970, following an explosion in an oxygen tank that severely damaged the spacecraft's electrical system, astronauts maintained air and water supplies through makeshift repairs and improvisation with equipment intended for use on the moon. Because the astronauts returned safely, despite failing to reach their goal, the flight has been described as a 'successful failure'.

Provisionality is not something to be overcome, a structural vulnerability, but rather the *underlying condition of the world*; it is part of being human in the world, and inherent in society as much as in the physical world. And, as Wentworth says, *we have the intelligence*. This moment, when relations between the ecological, geological, social and political are so densely intertwined, calls for experimentation in the improvisation of life. Improvisation is about resourcefulness, about seeing poetic possibilities and making creative responses to the moment. By exploring competing interpretations of our futures, we can begin to ask new questions about how we can use our intelligence. One of the things we want to encourage with the four conversations that are recorded in this publication is a sense that climate change opens up compelling new questions and invites ambitious and imaginative approaches to rethinking the world. Constructing for the unforeseen suggests a structured improvisation; like conversation, it is rigorous and practised and involves using our wits.

Four dialogues

History

These four discussions address a new field: 'culture and climate change'. This is a subject that considers, among other things, the response of artists, writers, performers, musicians, broadcasters and comedians (and, through them, the audiences that they reach) to the changing climatic conditions in the world and, more significantly, to the challenges that come from the new idea that we are largely responsible for these changes. As producers of the podcast, and editors of this volume, we wanted to start by asking where this subject – culture and climate change – originated and how far back it goes. We wanted to see how much it was part of other strands of political, anti-industrial or apocalyptic thinking and whether it has taken the place of other familiar and deep-rooted anxieties about the spread of germs, for instance, or nuclear proliferation. When did the subject of culture and climate change first emerge, who first engaged with it, how has it developed, and in what ways does it differ from the kinds of subjects that are closely related to it?

The first of the discussions took place in the imposing surroundings of the Examination Schools at Oxford University in September 2010. It was the opening morning of the fifth TippingPoint conference to be held in Oxford and the audience comprised 100 or so conference participants, including artists, scientists, writers and theatre directors. On the panel were Nigel Clark, senior lecturer in human geography, Siobhan Davies, choreographer, Wallace Heim, writer and editor on environment and performance, and Diana Liverman, professor in environmental policy. The discussion was chaired by Quentin Cooper, presenter of BBC Radio 4's *Material World*.

Quentin began by explaining to the audience that these discussions aimed to establish a framework for thinking about these cultural responses, for making them more engaging and more wide-reaching. This being the first of the four discussions, he wanted to discover what has been effective so far and what hasn't, what has raised awareness and what – if anything – has changed policy, practice and behaviour.

Quentin Cooper: Diana, can you recall when climate change first emerged as an area of research and when you first became aware of its consequences?

Diana Liverman We've been increasing the greenhouse gas concentrations in the atmosphere since the Industrial Revolution, but I would track the first interest in climate change to the Stockholm Conference in 1972. That's when I started to become aware of climate change as an issue. Then it sort of bubbles along. The really big news – at least in North America, where I've been based for a lot of my career – was that in 1988 there was a massive heat wave. At the same time, Jim Hansen, a famous climate modeller, got in front of the US Congress and said, 'We're warming the planet, I am absolutely sure'. That produced a lot of interest among scientists and policymakers and, in a way, led to the formation of the Intergovernmental Panel on Climate Change (IPCC) which produces reports telling us what's happening. In 1992, the UN Convention on Climate Change gave us a policy framework. So that period from 1988 to 1992 was really important. Along with that, the arts community, writers and artists started to get interested. But even from 1972, with the flowering of the environmental movement, artists start to be interested in the environment. Going back a very long way, artists have been important in relation to the climate. When I was an undergraduate I remember going to the Tate and seeing exhibitions of paintings by Turner and Constable and seeing how amazingly well they portrayed weather and climate. The history of climate change project at the University of East Anglia actually used the art of previous centuries to reconstruct what the climate was like.

Greenhouse gases act as a blanket over the earth, retaining heat. This natural phenomenon makes life on the planet possible, but human activity has increased levels of greenhouse gases, causing global temperatures to rise. The primary greenhouse gases in the earth's atmosphere are water vapour, carbon dioxide, methane, nitrous oxide and ozone.

–

The United Nations Conference on the Human Environment, held in Stockholm in June 1972. The declaration of the conference was the first document in international environmental law to recognise the right to a healthy environment.

–

The IPCC is an international grouping of over 300 of the world's best climate scientists charged with reviewing and reporting on the latest inter-national science, impacts and responses to climate change.

–

In 1822 John Constable (1776–1837) painted a series of 'landscapes' from his front garden in Hampstead, showing nothing but cloudy skies. On a number of these paintings, he noted the time, date and weather conditions. Evidence that Constable was aware of Luke Howard's classification of clouds came to light in 1972 (see John Thornes, 1979 and 1999). In 1995, the German artist Gerhard Lang photographed the sky from the same location, on the same dates and at the same times for his work 'John Constable's clouds are still passing'. See www.gerhardlang.com.

–

Hubert Lamb (1913–1997) founded the Climatic Research Unit in the School of Environmental Sciences at the University of East Anglia in 1972. In his book *Climate, History and the Modern World*, Lamb draws on paintings by Brueghel as documents of the Little Ice Age, and discusses how the sunset colours of Turner's paintings coincided with a sustained period of stratospheric dust from volcanic eruptions.

Wallace, can you explain how you came to be interested in thinking about art and ecology, and performance in the environment?

Wallace Heim I got started in this area because I fell in love, which is often how you get interested in things. You can become imbued with someone's ideas when you're in love with them. My late partner had worked with Gregory Bateson in California in the 1970s, on ideas about co-evolution and about mixing science, art and all of human culture. I was working at the time, in the 1980s and early 1990s, as a set designer in theatre and television. As I got more interested in Bateson's ideas and in things ecological, I realised there was nothing out there in theatre to see that reflected this. So I thought I'd change careers. I'd learn everything I could about ecological thinking and start to put together ideas about performance and ideas about ecology, almost like a 'thought experiment', to see what would happen. It's taken some time, but it's now not just a thought experiment, it's happening 'out there'.

Gregory Bateson (1904–1980) was an anthropologist, social scientist and linguist. His 1972 book *Steps to an Ecology of Mind* applies cybernetics (the study of regulatory systems, both physical and social) to the field of ecological anthropology. His book *Mind and Nature: A Necessary Unity* (1979) was a more philosophical work on the 'pattern that connects' in living systems, and the connections between culture, nature, science and the human.

Were you specifically interested in the climate change aspect or was it the broader questions of the environment and nature?

Wallace Human relations with nature is a very big area to consider. Climate change as a concept usurps or eclipses a lot of other environmental interests.

And you're particularly interested in taking performance arts out of their usual structures and into a more everyday context?

Wallace In the 1980s, art was being redefined: it

was moving out of the gallery or the theatre, it was engaging much more with activism and the role of the artist was being completely redefined. I think you get to a point as a practitioner, as an artist, where the conventions are not good enough. So what do you do? You have to change the conventions. This is what was happening in the 1980s. Some of that work did have very strong connections to people's relations to the environment or with nature.

Siobhan, what has been your involvement with the idea of climate change?

Siobhan Davies At about the same time that I was invited by Cape Farewell to go to the Arctic, I was building a studio in London. The idea behind the studio was that I would find a place that was both safe and exciting for dance artists, where they could discover more delicacy, more sophistication, more methods of exploring the physical, the emotional, the thought. And then I went to the Arctic and my body was nowhere. There was no such thing as safety. I knew I wasn't going to die because I was being looked after, but there was a moment in which I felt my life was suspended. The further I went from the ship, the more I felt that an umbilical cord was stretched. So there I was, in tension between trying to create 'safety' for the art form to move forward, into its potential, in London, while being thrown into this place of extremes where my body was in danger and completely unsophisticated. The outfit necessary for the Arctic, the layers of protective clothing, comes half way down your crotch and you can't move. The film director was saying: how would you move out here? And I hadn't got a clue.

Did you feel that you had to make an artistic response

In the 1980s and early 1990s, artists such as Mierle Laderman Ukeles, Alan Sonfist, Mel Chin and Lynne Hull were working with ecological themes of waste, remediation and habitat restoration in urban and rural environments. Performance artists, like Ana Mendieta, were working on the land with themes of gender and the body. Political and activist theatre was staged on the streets – for example, the large-scale spectacles created by Welfare State International in the UK.

–

Cape Farewell was founded by the artist David Buckland in 2001. The organisation brings artists, scientists and communicators together to stimulate the production of art founded in scientific research, describing climate change as 'a cultural challenge'. Since 2003, Cape Farewell has led eight expeditions to the Arctic, and one to the Peruvian Andes. Artists and writers on these expeditions include Ackroyd & Harvey, Anthony Gormley, Ian McEwan, Vikram Seth and Rachel Whiteread as well as contributors to this volume: Marcus Brigstocke, Quentin Cooper, Siobhan Davies, Beth Derbyshire, Charlie Kronick, Ruth Little, Vicky Long and Joe Smith. www.capefarewell.com.

–

The Siobhan Davies Studios in south London were designed in collaboration with Siobhan Davies and members of her company by Sarah Wigglesworth Architects. The practice specialises in low-energy and sustainable building, using readily available materials in an inventive way. The studios, created in a Victorian school building set in the playground of a functioning primary school, won a RIBA award in 2006. www.siobhandavies.com.

–

Director David Hinton shot footage on the first three Cape Farewell expeditions. The resulting film, *Art from a Changing Arctic*, co-produced by Cape Farewell and the BBC, was first broadcast February 2006.

that commented on climate change?

Siobhan I had to use my intuition. I would normally want to be completely thoughtful and completely accurate and none of my old tools for making work were available to me. So it was intuition that drove me to make the two pieces and I'm lucky that they worked. I think how I make work is different because of that experience. I can't necessarily produce a 'thing', the object which will help us transform, but rather a way of working: more connectivity, trying to work at an atomic level. In other words, if we think of each of us as an atom that will evolve, coalesce and, over a period of time, find different patterns and structures – that's what I work with.

The two pieces were *Walking Dance* (2005) and *Endangered Species* (2006). Siobhan Davies joined Cape Farewell's expedition to the Arctic in 2005. She created *Walking Dance* during the expedition, using the crew as her company. Her journal records: 'It is incredibly hard to think what kind of movement would have any relevance up here. Any of the full range of movements that I am able to articulate in a warm studio seem grossly over the top in a landscape such as this. The best movement is to walk or to run. Walking gets us somewhere and keeps us warm. It is a simple action, but there is an efficiency about it and in many layers of clothing it became denser than we had ever experienced. Dance here is then reduced to the familiarity, the straightforwardness, the elegance of the walk and the run.' (Buckland *et al.*, 2006) Following her return, Davies created *Endangered Species*. See note page 89 for details.

A lot of us talk about the 'power of the environment' from the comfort of our climate-controlled homes. We don't realise the rawness of what we are affecting. You're trying to get that across to us.

Siobhan One of the things that the arts can do is make us much more aware of our peripheral understanding, bring that back into the centre and start using that knowledge.

Nigel, do we need to take account of the cultural context in which climate change research appears?

Nigel Clark When I became a social scientist I was surprised at the limited range of materials and ingredients that social scientists worked with. I gradually started encountering artists, who worked with a much broader range of materials, of ingredients, who were allowed other things – whether it's pigments, or projections or living organisms – and allowed to do things, to make a difference. I found that made a huge difference to the way that I did social science. Just talking about culture and social processes wasn't quite enough. As a social scientist, I'm supposed to be the voice of the social and the cultural. But I keep on finding that I really want to be the voice of a dynamic planet of organisms that run around and take us by surprise. Working with and talking with artists has helped me as a social scientist.

You use the phrase 'dynamic planet': the earth has always changed, but there are changes happening on top of this which are caused by human activity. Do we need to reflect in arts and cultural interactions the differences between what's caused by us, what's happening anyway, what's within our control, and what's beyond our control?

Nigel Most social science works on about a 300- or 400-year timescale. Increasingly, conversing with earth scientists, I've got to get my head around things that happened on timescales of thousands or tens of thousands or even millions of years. That is the real challenge for us all, for artists, for social scientists. Earth scientists have got a headstart on that. They've got a long timeframe, which is incredibly important when thinking about the really important things that are happening right now – to think in the context of thousands of millions of years.

What is the earliest significant date for you in terms of human relations with nature?

Nigel I think the Lisbon earthquake in 1755 was one of the most frightening things that Europeans had ever experienced. It pretty much wiped out a major European city, right at the time when the

See Nigel Clark, *Inhuman Nature* (2010) which explores 'the volatile relationship between human life and the physical earth', considering, among other topics, the 1755 Lisbon earthquake, abrupt climate change, Hurricane Katrina and global mobility.

–

Between 10,000 and 100,000 people died in the Lisbon earthquake on the 1 November 1755. One reason for the event's impact on the popular imagination was that it struck on All Saints' Day, and many prominent churches were destroyed along with their congregations. The earthquake provoked responses from Enlightenment thinkers including Voltaire, Rousseau and Kant. Voltaire's *Candide* includes a fictionalised account of the earthquake, with the hero questioning the idea that a benevolent deity supervises 'the best of all possible worlds'; Rousseau suggests that the disaster was exacerbated by the fact of people living so closely packed in a city. In a 1931 radio broadcast for children, the German literary critic Walter Benjamin noted that 'no one was more fascinated by these remarkable events than the great German philosopher Kant [...] He eagerly collected all the reports of the earthquake that he could find, and the slim book he wrote about it probably represents the beginnings of scientific geography in Germany. And certainly the beginnings of seismology' (Benjamin, 1999: 538).

modern subject was new and fragile and nervous. A lot of thinking about what it is to be a modern subject, is repressing this phenomenal event. In some ways we are only just coming back to that moment.

The notion of the 'modern subject' is used to identify a shift in the way people have understood their place in the world, from the medieval sense that they were defined by their place in the cosmic order, to the modern idea of the self-defining individual.

How much of a role has culture played in the evolution of popular understanding of climate change? How much impact has it had on politicians, policymakers and others with the power to make a difference?

Diana Not much yet, particularly if I think of the American public. I think that their framing of climate change is deeply informed by the popular media and not the 'artistic media'. It's hard to see the influence of art and culture in politics and in individual human behaviour – and even the science that people know in relation to global warming is pretty minuscule. But I think that there are pockets of inspiration. There are moments when I see hundreds and thousands of people inspired by art to think more deeply about the problem of climate change, but when I actually think about the magnitude of the solution, the politicians that must make the decisions, the millions of individuals that must change their behaviour, I don't sense that the culture and the arts has really reached them yet.

Wallace Part of me wants to say that's the wrong question. Of course you want the arts to influence everything but in a sense that's asking for a very instrumental view of the arts. If you start with that premise you're going to get a very different kind of artistic and cultural response. You're going to ask for education, you're going to ask for impact, you're going to ask for all those things that policymakers can hold on to. I'd rather give policymakers something that they can't hold on to. I want to keep subtlety and nuance and complexity in there. I think we've learned that messages don't work. What's interesting is that some artists are now starting to work within the legal system, or work within the judicial system, or directly within the political system. I find that expansion of where artists can work very interesting. It's a rather more subversive way of working.

Siobhan One area is the gaps between disciplines: either between different artistic disciplines, or between the judicial, or the scientific, and the arts. If we are not too overly directed – in other words, telling scientists and artists that they *must* meet and they *must* make something – the tension between the exactitude of those two forms and what happens in the gap between them might be the energy plane that we work on. But we are not doing it yet and the urgency requires that we do.

Do we need to have some criteria to look at cultural responses to climate change and think about whether they are making a difference? Or is it just perfectly okay if a cultural response doesn't affect anyone at all?

Nigel Art has had a huge impact, but almost incidentally. Earlier in the twentieth century, artists did a lot of work breaking boundaries. That eventually fed into consumer culture, helped turn us into the sort of people we became in the second half of the twentieth century – the sort of people that confronted the climate change message. In all sorts of ways, the avant-garde arts prepared us for climate change. But early on there was a lot of talk about limits to growth, restrictions, belt-tightening, narrowing horizons. And the arts in the twentieth century had prepared us not to welcome those

The blurring of art and activism and the development of social practice arts have extended where and how artists work with social situations. Some artists are beginning to use Freedom of Information requests, judicial reviews and the lobbying of corporate shareholders and Ministers of Parliament for their research and activist potential, such as PLATFORM's work on the environmental record of the Royal Bank of Scotland. www.platformlondon.org. Some artists work more directly within established legal and political systems, often covertly. Raivo Puusemp was elected mayor of Rosendale, New York and transformed the governance and the fortunes of the town. Only five years later was it disclosed as an artwork. Using a different strategy, the Yes Men impersonate bureaucrats, infiltrate corporate situations and advertise the results of mock legal decisions and campaigns. Artist and activist John Jordan, a co-founder of Reclaim the Streets, set up the Clandestine Insurgent Rebel Clown Army and works with the Camp for Climate Action, bringing a creative approach to direct action.

–

The Limits to Growth was an influential report published by the Club of Rome, a think tank, in 1972: 'If the present growth trends … continue unchanged, the limits to growth on this planet will be reached sometime within the next hundred years. The most probable result will be a rather sudden and uncontrollable decline in both population and industrial capacity' (quoted in Hulme, 2009: 260).

things. We were geared up for something else. We didn't want to hear the message about restriction, limits, belt-tightening. It's taken a long time for climate change to bring about a new message that isn't about restricting limits. It's about things we can't know and unpredictable things. I think the arts and the sciences have converged again; they're coming together after the earlier period of being drawn apart.

This tension has been explored by David Buckland in describing the work of Cape Farewell: 'We intend to communicate through art works our understanding of the changing climate on a human scale, so that our individual lives can have meaning in what is a global problem.'

Do we need to think about climate change differently from other subjects? Does it have its own rules? Does it have its own challenges? Is it different, as subject matter, because of the magnitude, the intangibility, the time shifts?

Siobhan The human being is obviously central to climate change, but when we first start to think about it, we think in terms of place and landscape. So then where is the human being? Is it part of the romance of it rather than the devil of it? War, sex and death have 'romance' in them – using that word rather carefully. It's a human practice and a human problem. Then suddenly you've got a vast sense of place – and where do we fit in? That's the dilemma for me as an artist. Suddenly it feels very small if I make something, if I put a human being in a context, where the subject is vast. So for me it's the tension between the enormity and the single figure. And I'm finding that hard.

A landmark in cultural responses to climate change is the 2004 film, *The Day After Tomorrow*. The government's chief scientific advisor, Sir David King, said, 'I'll endorse this film on condition we have screenings at which there are scientists present who can give people the facts'. Is this a way forward –

to accept that the art may sometimes have to be compromised in order to reach a wide audience, but you can get to the science from it?

Diana I would say that Al Gore's film *An Inconvenient Truth* is far more important – even though it was quite deeply boring in places. For me, the most powerful moment was the end – the song – that made me weep. The problem with *The Day After Tomorrow* was that the science was impossible and it was overdramatised. I don't think the great climate change movie has been made. [And yet] there have been movies on other powerful social factors that change things. I do a lot of my work on how climate change will affect the developing world and the poor. Is anything connecting art and culture in terms of those millions of people who are going to be affected by climate change? Another issue, which I'm not sure that scientists and artists are addressing together, is the way in which climate change has become commodified – it's being sold through the market and carbon trading. That's taken it away from the deeply human responses that scientists and artists can have together.

How important is it to reach a mass audience through cultural artistic responses? Or is it more important to get to the policymakers, industry leaders? If one of them comes to a puppet show that happens to be about climate science, isn't that better than something like *Avatar* that vaguely permeates the consciousness of millions of people for a half an hour?

Wallace I wouldn't put it as either/or. We don't have very strong critical frameworks for looking at each within the context in which they happen. I think it's quite unfair to compare *The Day After Tomorrow* or

In *The Day After Tomorrow* (2004), a series of extreme weather events leads to a new ice age: the melting of polar ice disrupts the North Atlantic current; a tidal wave floods Manhattan; and snow and ice cover large parts of the northern hemisphere. A financial success, the film took over $500 million at the box office and over $100 million in DVD sales. Audience survey research in the US showed that the film 'had a significant impact on the climate change risk perceptions, conceptual models, behavioral intentions, policy priorities, and even voting intentions of moviegoers' (Leiserowitz, 2004: 34). A UK study of filmgoers came to similar conclusions, but also noted ambiguities people felt in distinguishing between science fact and fiction (Lowe *et al.*, 2006).

—

For Diana Liverman's recent publications on climate and development, see *Food Security and Global Environmental Change* (Ingram et als, 2010), and 'Copenhagen and the Governance of Adaptation' (Liverman and Billett, 2010). See also part III of Richardson, Steffen and Liverman (2011).

The Age Of Stupid with a puppet show. They're doing different things and I wouldn't want any of them not to be done. Underneath that is the question: does the science have to be right in order for this to be good and to be working? I think there are so many other things that have to be right too: the ethical quality, the political analysis, as well as the science. One area that hasn't been very well explored is the emotional qualities of these works. You get the big blockbusters, but the emotional quality is really banal. For me, that's as important as the science.

Nigel One of the things that *The Day after Tomorrow* did was to bring in the story of abrupt climate change – in a pretty dodgy way, but it put it on the agenda. There was that amazing image of the earth from space – fully glaciated. We'd had years of that image of earth seen from space – a static image. It did some wonderful things but it gave us a very static earth. *The Day After Tomorrow* brought us a different perspective – a glaciated earth. The whole earth had shifted. Okay, it shifted a little bit too quickly – it shifted overnight which was a little bit exaggerated – but we have been finding out for the last decade or so that climate change can happen very rapidly. It can happen in a lifetime. So the film was an exaggeration but it wasn't way off. Many people, human beings, hominids, have lived through very rapid changes in the past. I'm thinking of Siobhan's demand to bring it down to a personal level: a lot of our ancestors way back in the past had to live through abrupt climate change. If it wasn't for them making it through those episodes none of us would be here. In some respects, the fact that we are here at all is fairly unlikely.

Climate change is not unnatural; something that affects

Puppeteer Mark Down of Blind Summit Theatre writes: 'A puppet is any object that we bring to life with our imaginations. In Bunraku puppetry, three puppeteers work closely together to make very lifelike movement. In order to master the art the puppeteers listen to the puppet and enable it do what it wants to do. Allow it to live. By submitting to the puppet in this way the puppeteers fall naturally into collaboration. Perhaps the earth is our biggest puppet?' See www.atlas-id.org/blind-summit.

–

Climatologists have identified 48 major advances and retreats of glaciation over the last 2.6 million years, together with hundreds of smaller spikes of rapidly fluctuating climate change. Each one of these shifts would have had cataclysmic consequences for our human or hominid ancestors (Burroughs, 2005). When the climate cooled abruptly, sometimes in a little as a few years, the likely effects in many places would have been much less rainfall, high winds, dust storms, vast forest fires and crashing animal populations. Evolutionary psychologist William Calvin surmises that the shifts were often 'so large and so quick that a single generation gets caught, forced to innovate on the spot'. He adds, 'Our ancestors lived through hundreds of such episodes – but each became a population bottleneck, one that eliminated most of their relatives. We are the improbable descendants of those who survived – and later thrived' (Calvin, 2002: 17, 3).

a species on a remarkable scale is not unnatural.

Diana This is different, because this is an ethical question; we are creating this change, which we didn't in the past. We have to take more responsibility than just reacting. And some people are much more vulnerable than others, including many people that didn't cause it. So whilst the experience of the past may give us some guidelines about how to adapt and how we might survive, the deeper ethical, emotional issue is that this is something we created. We can't blame it on outside forces.

Do we need to get down to the nitty-gritty of the science? Is it important that scientists and artists work closely together?

Diana We need in some cases to get down to the nitty-gritty. Take two popular solutions to climate change: carbon offsetting and geo-engineering, both of which we can get terribly wrong. Both the artistic and scientific communities could have something to think and say about both issues, whether it's artists worried about flying around the world and whether they should offset their flights or whether it's engaging with artists to imagine a future where we geo-engineer the planet in order to cope with climate change. That's one of the new areas worth having a conversation about. We may get to the point where we are not going to do anything to reduce emissions so we've got to start putting gases and sulphur dioxide into space. What are the impacts of that? What's the meaning of that?

Siobhan Artists and the art movement can help change attitudes through human understanding and

International lawyer Michael Byers has written: 'Governments that today refuse to prevent climate change may well come to be regarded in the future as having perpetrated international crimes' (Byers, 2005).

—

Carbon offsetting: a system of purchasing a certified reduction in carbon dioxide emissions somewhere in the world to balance an equivalent expenditure. 'The first voluntary offset deal was brokered in 1989 and the market has grown exponentially in the last few years. In 2007 there were over 60 different organisations in the UK alone offering such voluntary carbon offsets.' (Hulme, 2009: 301)

—

Geo-engineering: a scheme to manipulate the physical functioning of the earth's climate system. Ideas include injecting sulphur into the stratosphere; sucking up biologically productive cold water from the deep oceans to stimulate growth of algae; building carbon dioxide extraction machines; installing giant mirrors in space; and fertilising the oceans with iron. Mike Hulme writes that geo-engineering 'raises novel ethical dilemmas for climate governance. Many ethicists would argue that deliberately seeking to alter the world's thermostat, with consequences that are not fully predictable, is a categorically different ethical judgement than seeking to reverse or contain the consequences of what, historically, has been an inadvertent human modification of global climate' (Hulme 2009: 317). See also 'Smoke and Mirrors: Some Reflections on the Science and Politics of Geoengineering' by David Humphreys, 2011, and a lecture given as part of Princeton University's 'Ethics and Climate Change' series by Robert Socolow: http://web.princeton.edu/sites/pei/ECC/fall08/socolow.htm.

human empathy and human engagement. We have to be open to receive what science is telling us. I feel that's what art knowledge can do: we change or adapt through some kind of receivership, and by being open and receiving what science is telling us, we may be able to alter how we respond to what is happening to us.

Is it important to think about whether we are making sustainable art? Or is it what is presented to the audience that is paramount?

Siobhan It's got to be both. If I can't produce an object [that transforms attitudes] then the methods I use to make work, or trying not to expend energy through touring, are the things I can alter, so I'm going to try and work on that. Or to work on connectivity: my community seems so much larger and more exciting now because I've tried to think horizontally rather than hierarchically. If you are thinking hierarchically, then maybe you are always trying to make the object. But if you are thinking horizontally, then you are trying to link with different people and different practices and that has felt more inventive.

The Furtherfield Media Arts Ecology programme raised the question: What are artists' responsibilities in regards to their impact on climate change? Ruth Catlow and Marc Garrett took inspiration from Gustav Metzger's Reduce Art Flights campaign and decided to stop flying for six months. 'We will not take an aeroplane for the sake of art. For the next six months we will find other ways to visit and participate in exhibitions, fairs, conferences, meetings, residencies. We will not fly for inspiration, nor to appreciate, buy or sell art.' See www.furtherfield.org and www.pledgebank.com/wewontflyforart. The poets Melanie Challenger and John Kinsella gave up flying, and therefore the chance of meeting, and wrote each other poems instead: www.artsandecology.org.uk/magazine/features/dialogue-between-the-body-and-the-soul.

Do we have to think about human-induced climate change versus non-human-induced? For a lot of people, the story is simply: climate change is a bad thing; it is to a certain extent our responsibility; there are things we can do to mitigate it. The question of how much is caused by us is irrelevant.

Nigel We are able to speak about what we've done and what the planet does in the same breath. It's not an either/or thing. The reason we can have such a massive impact on the planet is because it's precarious. Its climate systems are poised. If they weren't poised and precarious, if they weren't capable of tipping and shifting, then we wouldn't have the impact we do. It comes back to Diana's point about responsibility. I'm thinking particularly about the different kinds of responsibility that artists and policymakers have. In a sense, they are different kinds of responsibility. One of the things that artists do as part of their response to the world is to experiment with materials and different ways of bringing things together. But it's an open-ended process, whereas policymakers have to deal with millions of lives and they have to have a sense of the outcomes of their policies. I think there is a different sense of responsibility.

People keep emphasising that climate change has no single cause; there is no single solution. Do we need to take the same approach with artistic responses? By any means possible: direct, didactic, subtle, intangible? Or should we attempt to focus our attentions on particular areas where we seem to gain most ground?

Wallace I think there are infinite ways of expressing things. There is infinite room for redeveloping and re-imaging and coming up with new forms of art. We've been talking about 'art and science'; there's a lot of other people involved as well. It's not just artists and scientists. Developing strong critical frameworks is a way of not just focusing on the art but focusing on the fact that the artwork is also made in its relation with its audience. It's not just the object. It's the relation it creates within the public realm. I think there is much more scope for thinking about that as well as just thinking about the particular artwork. Where does it fit? How does it work? Who does it speak to? And each form is going to come up with something different.

Diana The research that I've done looking at how cities, states and individuals are responding to climate change suggests that yes, you want a lot of different actions, but networks make a massive difference. Networks have the power to link people together, to share best practice, to inspire each other and to influence policy and the public. However diverse the artistic response, individuals linked into networks is what could make a difference.

Nigel I want to come back to the unknowingness of doing art. I come from a part of the world where introduced organisms transformed entire continents. Rabbits got let loose and look what they did to Australia and New Zealand. They completely transformed the landscape. Similarly, we just do not know where a lot of aesthetic practices are going to go and what they're going to lead to. Thinking about geo-engineering – who knows that there is not going to be a coming together of artists, scientists, policymakers to start experimenting, in an aesthetic and a scientific sense, with what the earth can do? That may be the sort of art and science collaboration that happens in the future.

Siobhan It's to do with the language in which what we are doing is made common to a larger group of people. I think we need more commentary – people looking at this and making comment on it and giving language to people who might not be involved in the arts and science, a way of feeling included.

Is there is one lesson from the history of cultural responses to climate change that can usefully inform what we do now and in the future?

Diana Liverman summarises the key issues in her essay 'Informing Adaptation', written for the World Resources Institute (WRI) www.worldresourcesreport.org/responses/informing-adaptation.

—

For example, the Transition Network was set up to support transition towns asking the question: 'How can we make our community stronger and happier as we deal with the impacts of peak oil and economic contraction while at the same time urgently reducing CO_2 emissions?' www.transitionnetwork.org; the C40 global network of city mayors was established with the recognition that cities 'have a central role to play in tackling climate change'. See www.c40cities.org. At COP15 in Copenhagen, December 2009, even as negotiations at the national level were faltering, networks of city mayors, young activists, artists and researchers were meeting and swapping ideas.

—

Nigel Clark has written on this topic, 'Either introduced intentionally, or smuggling themselves amidst licit cargo, hundreds of species from the temperate north traversed the oceans to the "new worlds" of the south [...] Cattle, let loose on the Argentine pampas to forage, proliferated into free-ranging herds so vast that they could not be counted, let alone corralled; a handful of rabbits released in Australia soon numbered hundreds of millions; dozens of varieties of European plants, imported to New Zealand as fodder, garden or hedgerow plants jumped the fence and ran rampant over the countryside' (Clark, 2003: 163-4). See also *Ecological Imperialism* (Crosby, 1986), and *They All Ran Wild: The Story of Pests on the Land in Australia* (Rolls, 1969).

Nigel A lot of culture – going back over tens, hundreds, thousands of years – is a response to climate change. It's how we got through climate change in the first place. The massive lesson of the last decade or so is just how uncertain the planet can be. We've responded before and we are going to have to respond again quite quickly.

Siobhan I've been looking at how the brain works, at studies on neurological pathways. We have quick hard drives to things like anger, jealousy, sex and territory, but we do have empathy and compassion and thought and feeling within us, and we don't expose ourselves to that. I'm wondering if we can use that practice to develop our thinking and our responses in a different way.

Wallace Aristotle came up with the idea of *phronesis*, which is ethical knowledge in a very particular situation: it's knowing what that situation calls for in terms of right action when you don't know what right action would be. When you've got no idea. No beliefs, no principles, no precedent. It's a description of how you respond in that situation which Aristotle went on at great length about. *Phronesis* is reason that shows itself in the ability to improvise.

Diana I think it's probably a lesson about seizing the moment. When I get desperate about responses to climate change I think it's going to take disasters to change it, which I don't want but I think will happen. At that point there are ways in which we can respond that could change things, whether we are artists or scientists. So it's a question of being ready for those moments that will happen over the next 20 years.

Responsiveness to change, adaptability and resilience are increasingly prominent themes among policymakers, replacing the 'sustainability' and 'sustainable development' discourses that have dominated since the late 1980s.

—

Siobhan Davies' reading includes *Rewriting the Soul* by Ian Hacking (1995) and *Understanding Emotions* by Keith Oatley, Dacher Keltner and Jennifer Jenkins (2006). The work of neuroscientist Antonio Damasio, speculating on the relations between emotions, rational thinking and the body, has also been widely discussed among performing artists. See, for example, *Descartes' Error: Emotion, Reason and the Human Brain* (1994) and *The Feeling of What Happens: Body and Emotion in the Making of Consciousness* (2001).

—

Phronesis is usually glossed as 'practical wisdom' or 'prudence'. It is described by Gerald Bruns as 'a condition of moral knowledge at the level of particular situations…knowing what a situation calls for in the way of right action, even when the situation is so complex and unprecedented that one experiences the shortfall of one's principles, beliefs, or patterns of conduct, or even one's sense of how things should go if they are to go right… *Phronesis* is … reason that shows itself in timeliness, improvisation, and a gift for nuance rather than in the rigorous duplication of results' (quoted in Heim, 2003: 195).

—

See Diana Liverman's conference report, 'Mind the gap. Policymakers must aim to avoid a 2 °C temperature rise, but plan to adapt to 4 °C' (New, 2009).

Publics

If we consider the ways in which artists – and, for our purposes, that includes writers, performers, musicians, broadcasters and comedians – have responded to climate change we must also consider how each of these groups interacts with its own audience or 'public'. The audience at an art gallery has a different set of expectations to an audience sitting at home watching a sitcom. These audiences also have varying levels of engagement and expertise in the subject and bring their own values and predispositions to the works. How does that influence what is produced? How much does the point of view of the audience (what they know, where they stand on this issue) impact on what is presented? What if some people simply don't want to hear any more about it? And has a shift taken place over the last two decades in what can be said and the way in which it can be said?

This second discussion took place in the middle of the temperate biome at the Eden Project in Cornwall. It was a bright, warm October morning (even more so within the biome) and the audience, consisting mainly of students and teachers, wore t-shirts and sunglasses. There have been more than 10 million visitors to the Eden Project since it opened, and some of the Eden visitors, who happened to be passing through the biome, also paused to listen. On the panel were Marcus Brigstocke, comedian, Vicky Long, producer, Tim Smit, entrepreneur and co-founder of the Eden Project, and Joe Smith, senior lecturer in environment and co-author of *Climate Change: From Science to Sustainability*. The discussion was chaired by Quentin Cooper, presenter of BBC Radio 4's *Material World*.

Quentin began by quoting the earliest reference he had come across to climate change in a piece of work that had reached a mass audience. The moment occurs in the 1967 film *Quatermass and the Pit*, when Professor Quatermass asks his archaeologist friend, 'If we found out that our earth was doomed, say by climate changes, what would we do about it?' The archaeologist replies, 'Nothing, just go on squabbling as usual'.

Quentin Cooper: The Eden Project is coming up for its tenth anniversary and is now a beacon of sustainability and greenness but, presumably, when you were planning it, all that was much lower down in the public consciousness?

Tim Smit We came at it from the point of view of the environment in the widest sense and wanting to connect people to the realisation that they were dependent, or rather interdependent, on the natural world. We wanted to be the world's first rock and roll scientific foundation because we were fed up with hearing an awful lot of intelligent people talking to a lot of other intelligent people about their own prejudices rather than addressing the people who weren't interested in environmental issues.

Something like 40 per cent of the British public still claim to be sceptical about climate change. Do you have more chance of reaching them through something like Eden than though a 'climate change play' or a TV programme which might select an audience already predisposed to be interested in those issues?

Tim We deconstruct 'climate change' and break it into its constituent building blocks, so that people feel they have some agency, are able to effect a change. I don't think any of us feel we can do anything about it as an umbrella title but we can do a heck of a lot if we have a new narrative which is about putting together the little bits that build up the constituent parts. Lots of people who talk about climate change bore the hell out of me. You get this hair shirt thing: you're going to have to give up this or give up that. I think the idea is to show people that life can be lived joyously and fully and that the challenges that are being thrown at us for the first time are worth facing. It's rather fun. It should be viewed as fun rather than, 'Oh my God, we're doomed'.

Is this a problem that's got worse? The more people try to draw attention to climate change, its consequences and what we need to do, with statistics and scientific research, the more they are seen as doom-mongers, hair shirt purveyors, frighteners, paralysers, rather than potential agents for change?

Joe Smith Since the 1960s, we have gone from climate change being a seminar room conversation among a few dozen people to the idea that the things we do every day affect the global climate. That's been accepted by 60 to 70 per cent of the population in the developed world – it's an amazing achievement, particularly of the science and policy community, and specialist journalists, who've told the story pretty effectively. I agree with Tim that we have tended to tell a story about our response to this in terms of fear and denial. The 40 per cent of people that we need to draw with us to build a working majority didn't arrive at the beginning of the twenty-first century expecting to be rationed. When you begin to move out of developed world countries and have international negotiations, you've got a lot of people in the South who think that fridges, access to cars and so on is part of the plot for them in the twenty-first century. So the fear and personal denial story is not going to build substantial action.

How do we get round this problem? If you have a TV programme about the environment, a play about climate change, or an article about sustainability, a lot of the people you most want to reach will turn the channel, turn the page, turn their mind away from it.

A number of campaigning organisations now specialise in reframing the terms of the discussion to avoid the language of 'doom'. Global Cool, for example, describes itself as a 'green lifestyle organisation' and works with celebrities and entertainment to 'show you how to live a greener life without sacrificing the things you enjoy' www.globalcool.org. The sustainability communications agency Futerra emphasises its use of 'evidence-based research on how best to communicate to change attitudes and behaviours' on issues 'from green to ethical, climate change to corporate responsibility'. www.futerra.co.uk.

–

There are several sources of quantitative polling data about attitudes to environmental issues and climate change: IPSOS MORI have long running polls www.ipsos-mori.com and Globescan conducts a regular Climate Change Monitor in 20 countries www.globescan.com/ccm_overview.htm. The representation of climate change in news media is explored in several books and many articles: Smith (2000) is unusual in including journalists and science and policy practitioner accounts; the contributions to Boyce and Lewis (2009) tend to take a critical line on media performance from an environmental NGO perspective. Boykoff (2011) brings together ten years of data in the most systematic review of news media performance, and Anders Hansen (2009) considers news media in the wider context of communications research, including popular media and advertising. Both Boykoff and Hansen offer comprehensive references to the academic literature.

Marcus Brigstocke I could have written a two-hour stand-up show about climate change quite easily by now but there is absolutely no point because the only people who would come and see it already agree with me. So the approach I've taken is to drip feed it into everything that I do, whenever I'm on the radio or doing a stand-up show on any subject, to try and keep it in there just a little bit. People are on to me, it's no sleight of hand – they know what to expect when I appear. In terms of creating comedy one of the easiest routes has been to mock the people who think that it's not happening, because I find them easily mock-able. They will say a great deal but when questioned they haven't read anything. For me, what's been more challenging has been to take the idea of sustainability and 'green living' – for want of a better expression – and express the positive in it, and how much I've enjoyed it.

See, for example, Marcus Brigstocke's appearances on the BBC Radio 4 satirical sketch review of the week's news, *The Now Show*, and his journalism, including this piece on climate change contrarians for *The Guardian* www.guardian.co.uk/environment/2009/feb/10/climate-change-marcus-brigstocke

–

How can popular culture take on sustainability, environmental and green issues and reach an audience that isn't already persuaded?

Vicky Long Imagination has a big role to play. We need to grab the public's imagination on this subject, and artists have a role to play in that. Yann Martel, the author of *Life of Pi*, has talked about climate change being an impersonal force that's deeply disempowering. In contrast, art and the making of art is personally involving for the artist and for the spectator. In working with artists and making exhibitions, films, events, working with places like the Eden Project, we've been able to bring people in, involve them in the subject and reach their bloodstream in some way. It's been very exciting. I see all the work I've done over the last five years as an experiment. I don't think there is any magic

Over four years, Yann Martel sent Stephen Harper, the Canadian Prime Minister, a book every two weeks. The books were chosen as being ones that 'expand stillness'. The 52nd book was the Cape Farewell publication *Burning Ice: Art & Climate Change* (Buckland *et al.*, 2006). Martel's covering letter describes climate change as 'an impersonal force, deeply disempowering' while 'art inspired by climate change, because the making of art is personally involving, a whole-person activity, is empowering, both for the maker and the spectator'. His letter to the Prime Minister continues: 'One thing is certain: our response to climate change cannot be purely political. Politicians have been dragging their feet – you among them – because of the power of the carbon-fuel industrial complex. It is citizens who must move first, and art is an ideal way to help them do that. Art wrestles with its subject matter on a level that the individual, the man, woman, teenager and child on the street, can engage with and react to. Once citizens are involved in the vital issue of climate change, politicians will have to follow their lead. See www.whatisstephenharperreading.ca.

answer to what works and what doesn't. But the experiment has been very vital and interesting and I think there have been many successes along the way. It's really about bringing people together and involving them. If, for example, at the Eden Project we have a comic, a musician, a scientist, an installation, a piece of visual work, I think that really works because there is something that just about everybody can relate to.

Working in collaboration, Cape Farewell and the Eden Project have presented installations (including Max Eastley's *Ice Clock*, Clare Twomey's *Blossom* and Beth Derbyshire's *Anthem*), participatory events such as the site-wide *Unplugged* music project with composer and musical director Tim Sutton, and a series of summer events offering the chance to engage with climate experts, stand-up by Marcus Brigstocke, and music by artists including beatboxer Shlomo and the songwriters and performers Liam Frost and Robyn Hitchcock. www.edenproject.com/whats-it-all-about/people-and-learning/music-and-art/climate-change-arts-programme.php.

Tim What we are talking about in terms of culture is permission to act in a different way, to create a mandate for politicians, public servants, individuals, to get behind a new vision of what the future is. That sounds pretentious – but I believe that we are living at a time which a hundred years from now could be seen as the dawning of something as important as the Renaissance. If we buy into the fact that we can take agency over our futures rather than [being] passive flotsam and jetsam on the top of it. Just dream a little. Say the current government decides it's going to radically transform the budgets and – with singlemindedness, with the urgency of a war effort – become completely energy independent, based on renewables. Because of the way that everything is monetised in our world intellectually, the only way they could get a mandate for that would be not simply as a response to climate change, it would have to be that we as a tribe get advantage through this brave, bold effort to show that we have understood the rhythms of the earth and are responding to them. That's why art has such an incredibly powerful role to play, to create, if you like, the wallpaper around the rooms in which the thinking is done, to give people the bravery to make that leap – because it feels the right thing. It needs to project onto those who are opinion-makers and decision-makers the sense that they are at a moment in history where if they do not have the bravery to make decisions they will be found wanting. You are living at a time when you could harness something brilliant in us which is aspirational, that will say something about who we are right now. That's the way to go. Not, 'We're doomed, we're polluting the oceans'. Whether that's true or not, that's not an inspiration for change. The idea that through your own agency and linking up with others you suddenly become tens, hundreds, thousands, millions of people. One of the beauties of modern communication is that you can create an atmosphere and persuasive proposition to a huge audience in a very short period of time.

Can you give me some examples of things that you think have worked?

Marcus It's a question of what you are willing to define as art and culture. They've got these bikes to rent in London – not free, but very cheap. People cycling around London on bikes sponsored by Barclays Bank is not necessarily art but it's altogether visually more appealing than people sitting, swearing and sweating in their cars. I live in London where people live stacked up on top of each other and actually I see a lot of very positive changes there. I see more and more people on bicycles – though not perhaps as many as drop their children off at school in a Sherman Tank. Simple things like the pedestrianisation of Trafalgar Square and various other bits of London where people are walking around. I think there are a lot of successes and a lot of examples where you can feel hugely encouraged.

Joe It's an incredibly ambitious intellectual question:

Literary critic Raymond Williams describes 'culture' as 'one of the two or three most complicated words in the English language.' In his book *Keywords*, he discusses three modern senses of the word: (i) 'a general process of intellectual, spiritual and aesthetic development', (ii) 'a particular way of life, whether of a people, a period or a group' and (iii) 'the works and practices of intellectual and especially artistic activity'. This last usage now seems the most widespread: 'culture is music, literature, painting and sculpture, theatre and film.' For Williams, it is 'the range and overlap of meanings that is significant. The complex of senses indicates a complex argument about the relations between general human development and a particular way of life, and between both and the works and practices of art and intelligence'. See also Clifford Geertz page 77 this volume.

–

Barclays Cycle Hire, a Transport for London scheme, was launched in 2010 with 5,000 bicycles and 315 docking stations in central London. After paying a £1 daily access fee, journeys of up to 30 minutes are free.

–

Long discussed by London planners and politicians, the pedestrianisation of a large portion of Trafalgar Square, including a four lane road in front of the National Gallery, was eventually achieved in July 2003.

how do we make sense of human beings' place in the world and their possible influence on the atmosphere? Almost all of the people who have any claim to authoritative knowledge on those questions agree that humans are changing the climate, almost certainly in hazardous ways. But we have to be slightly more honest about that being a long and uncertain journey intellectually, and try to communicate that [uncertainty] in popular culture. I'll give you two examples of influential films; one I view as a hit, and the other as a miss. Al Gore's *An Inconvenient Truth* spends 158 minutes bashing you over the head with very assertive phrases like, 'The science is settled', and then spends two minutes with a lot of Second World War references along the lines of 'Come on guys, we can do it together.' You know, 'We will change six billion light bulbs' and so on. I think that set up a whole set of hazards for the environment policy and science community, particularly the political community. Contrast the Al Gore film with a pair of documentaries made by the BBC at about the same time. They were fronted by David Attenborough, universally loved and trusted. He presented himself as having taken a journey with the topic and arrived at a conclusion: it's clear that on the balance of evidence we should act. The second really important thing was they threw lots of money at design and worked hard to bring to life the idea that the carbon we generate in everyday life is tangible. They made graphic carbon blocks appear above the household of a typical American family. In the second part it was all about how you would take those carbon blocks out of the sky without damaging the quality of life of that American family. So there was an even balance, where Al Gore had offered a 158 minutes of 'grimathon'. The second film had a much more even balance, a much more compelling

The website linked to *An Inconvenient Truth* offers printable tips designed to foster individual energy efficiency at home, when commuting and on a national scale. www.climatecrisis.net

—

Are We Changing Planet Earth? was shown on BBC One on 24 May 2006, and *Can We Save Planet Earth?* on 1 June 2006. A DVD which includes both documentaries, *The Truth About Climate Change*, was released in 2008. Attenborough's opening words are: 'This is our planet: planet Earth. It contains an astonishing variety of landscapes and climates. Since life began, around 4,000 million years ago, it has gone through extraordinary changes in its climate and in the species that live on it. But now it seems that our planet is being transformed – not by natural events, but by the actions of one species: mankind.'

case. I want to push much harder at questions like – and this is something that comedy and drama as well as factual people can do – what have we won from 60 years of a really hearty carbon burn? What have we won in terms of relationship breakdown, relations between generations, the quality of life in a city or in a town or in a village? We know instinctively that we've pumped all this stuff out of the ground and into the atmosphere and actually won very little in terms of advances in quality of life. So it's not saying, 'Right, what do we have to give up when we turn off the carbon tap?' It's rather saying 'Right, as we get smart about energy, how can we do a better job of winning a better quality of life?'

This argument is developed in *Do Good Lives Have to Cost the Earth?* (Simms and Smith, 2008). Among the recent publications which have questioned the notion of 'quality of life' as it is currently understood in most anglophone industrialised countries are *Affluenza* by Oliver James (2007), *The Spirit Level: Why More Equal Societies Almost Always Do Better* by Richard Wilkinson and Kate Pickett (2009), *No Logo* by Naomi Klein (2000), *Amusing Ourselves to Death: Public Discourse in the Age of Show Business* by Neil Postman (1985), *How to Be Idle* by Tom Hodgkinson (2005) and *Last Child in the Woods: Saving Our Children from Nature-Deficit Disorder* by Richard Louv (2005).

The David Attenborough documentaries were the first ones to be shown in peak time in the United States that dealt with environmental issues. When Attenborough's previous series *Blue Planet* was shown in the United States, they broadcast all the nice, pretty episodes about the animals and dumped the final programme which dealt with the environmental issues. Is that the difficulty with the mainstream?

Marcus Looking at the pretty animals and fish on *Blue Planet* and *Life on Earth* – that's the positive message, trying to reconnect people with the fairly small planet that we all share. That's the good stuff. I don't think they're deliberately oblique, I think they do a better job than anything 'explaining', because you can't really talk about climate change without touching on the idea that this is very important and very urgent because life as we know it will change very, very dramatically. So to try and engage people in more creative ways in the solutions is better. I think there is an appetite for that. It's bizarre how sustainability, climate change, all the rest of it, was a big political subject until the economic crash and then it disappeared. And it's not really re-emerged as part of the narrative of what politics is. It's just gone. It's extraordinary that it was something we 'did' for a little while: 'Oh, we all gave ourselves a bit of a scare where that's concerned'. It often comes up when I get into a battle with someone who hasn't read enough about it. They will say, 'Well, I mean the last one was that hole in the ozone layer wasn't it, and they said that was going to be awful and then it went away'. Well, if it has gone away, isn't it because we acted on it? There was a problem. We found a solution. We did it. And it seems to have worked.

Part of being a comedian is being nimble enough that you don't deliver the same material night after night: you adapt what you do, feel who you are talking to. Do we need to do that when we are talking about climate change and sustainability – pitch it differently to different sections of the public? Is that a lesson that all the art forms can learn from comedy?

Marcus I tweak it a lot. I talk in my shows about lots of subjects. Theology is a good one – you'd think that would be very divisive and upset people. Not nearly as much as talking about climate change does. My audience divides into three groups at that point: 'It's not happening at all and even if it was there's nothing we can do'; 'It's definitely happening, it's all our fault and we feel ghastly'; and then everybody in between. None of those groups are particularly comfortable about laughing about climate change, the environment, sustainability, whatever you want to call it, which is exactly why I am perversely interested in it. I enjoy that. Dealing with that thing where people feel exposed and feel that they are being asked to do something. A simple icebreaker

The economic crisis which began in December 2007 with the sudden losses from sub-prime loans, increased dramatically in severity in September 2008 with the collapse of Lehmann Brothers, leading to panic in the banking sector. The underlying causes of the crisis range from a bubble in real estate and equities to the steep rise in oil and food prices. James Painter's comparison of the media presence at the UN FCCC COP 15 in Copenhagen and COP16 a year later in Cancun plots a dramatic falling off of coverage http://politicsinspires.org/2011/01/media-interest-in-climate-change-takes-a-dive/. Max Boykoff and Maria Mansfield have plotted world newspaper coverage of climate change/global warming since 2004 and demonstrate a dramatic spike and falling away of coverage across 2009-2010: http://sciencepolicy.colorado.edu/media_coverage/

–

Environmental campaigner Chris Rose and marketing specialist Pat Dade propose a similar approach to expanding engagement in climate change by focusing on the distinct values and motivations of different groups in society through a 'values modes' analysis. They divide people into three groups according to their values: 'settlers', 'prospectors' and 'pioneers'. See, for example, Rose, 2010.

that I nearly always do when I talk about it is to say, 'Who here has got eco bulbs in their house?' and most people put their hands up, some of them very quickly because they want others to know that they've got eco bulbs in their house. Then I say, 'Good, so like me, you spend the first ten minutes in every room in complete darkness'. I acknowledge the fact that some of these things are – not difficult or hard, but just a bit weird, and they take a little bit of getting used to.

Is there a difficulty in being perceived as an artist too closely associated with climate change? Vicky, you work with some artists who willingly embrace that term. Is that difficult, or are they happy to say, 'This is the biggest issue facing our planet today? I'm an artist. I have no choice but to try and address it through my work.'

Vicky Cape Farewell works mainly with artists who haven't worked on climate change. We bring them into contact with climate science, mostly on a journey. We take artists and scientists together on a boat to the Arctic but we've also visited the Andes in Peru and trekked there. That gives them a very direct route into the science of climate change. People change on those journeys. There's something about an environment in which people can come together, talk, share ideas, hatch new ideas, that changes somebody and changes what they want to do with their practice – and where they are in the world. And there is also something very important about giving people a story to tell. It's not just about the work that you do, but the story that comes with that. I think that good ideas and work travel on the wings of a good story.

Tim The BBC rules for 'equality of access' have created the most incredibly beige betrayal of our culture. Regardless of the fact that a million people know the world is not flat, if one person thinks it is flat, we've got to give equal balance. I'm not singling out the BBC, but there is a huge lack of courage. Do you know that Bernard Shaw line from *Man and Superman*? 'The reasonable man adapts himself to the world; the unreasonable one persists in trying to adapt the world to himself. Therefore all progress depends on the unreasonable man.'

We've just heard about the problems of the media notion of balance, that if a thousand scientists think there is overwhelming evidence for climate change and you can find one that says 'I'm not so sure' then that is reported in that way. One consequence of this story moving from the periphery into the mainstream is that it will get that kind of treatment. Do we just have to accept that is the way the media works?

Joe For a couple of years, in the run up to Copenhagen, editors and journalists followed a line broadly accepted in the science policy community – that climate change is a problem and humans must address it. That built up like an infection in editors and journalists. They resented the sense of being enrolled in some kind of soft left agenda because of the nature of the things you have to do if you take climate change seriously. You're enrolled into a body of things that do imply action. So I think they were looking for an opportunity to lance it. There is still a sense of 'Climate change? We're through that. We're on to another story now.' But there's a difference between climate and weather! This is intellectual weather that we've just suffered; the long-term signals are the thing to focus on. And the

Copenhagen: the United Nations Climate Change Conference, held in Copenhagen, Denmark in December 2009. The conference included the 15th Conference of the Parties (COP15) to the United Nations Framework Convention on Climate Change and the 5th Meeting of the Parties (MOP5) to the Kyoto Protocol. The Copenhagen Accord, produced at the 2009 conference, recognised that climate change is one of the greatest challenges of today and that actions should be taken to keep temperature increases below 2°C. It was 'taken note of' but not 'adopted' by most of the participating countries.

—

The difference between climate and weather featured in a television sketch by comedy act Armstrong & Miller. See www.ashdenawards.org/blog/Armstrong-miller-climate-change.

long-term signals are that we are bedding down an acknowledgement in the wider culture but also in news culture that climate change is a serious issue that we will have to face at some time. I'm surprised and deeply disappointed that we didn't view the economic crisis as an opportunity to rethink the economy in ways that would add the environment into our accounts. We've missed the boat and that's a measure of something very important which is that popular culture around environmental change lacks a political culture. We need to make space for a really vigorous debate but the one thing we all have to agree on is we must account for biodiversity in our household and national accounts. We must give carbon a price that allows us to be energy intelligent. How do we make more space for a vigorous political culture around these issues? Mostly that's going to happen in the factual realm, in news. But, actually, we didn't work out some of the trickier social questions of the late twentieth century in those spaces. Comedy and drama were really important to how families were rethought. Ethnicity, gender – some really big topics in society – didn't happen in the news or factual space. Comedy and drama helped us work it out and then we worked it out at kitchen tables and in pubs.

The term 'biodiversity' is a contraction of 'biological diversity' and describes the variety of life on earth and specifically the total sum of the genes, species, habitats and ecosystems in a given environment. A major international initiative convened by the UN Environment Programme on The Economics of Ecosystems and Biodiversity has published a series of reports on this theme, including a summary report: *Mainstreaming the Economics of Nature* (2010): www.teebweb.org/TEEBSynthesisReport/tabid/29410/Default.aspx.

–

This was a central concern of the Stern Review on the Economics of Climate Change, a comprehensive review of the economics of climate change, carried out by Sir Nicholas Stern in 2005. The main conclusion is that the benefits of strong, early action on climate change considerably outweigh the costs. (See chapter 6 for discussion of economic modelling of climate change impacts and chapter 10 for analysis of the costs of action on climate change, Stern Review, 2006).

Trying to get the mainstream media to engage with issues around climate change is complicated. But for the climate deniers it sometimes seems much simpler. They can make little mocking comments. They can say you don't need to do anything. They can be very reassuring. They can get the quick laugh. Is the game skewed in their favour?

Marcus *Top Gear* is the most successful programme that the BBC makes, which genuinely makes me want to hang myself from the top of this biome as a protest. But it's immensely successful and it thrives on exactly that: Jeremy Clarkson looking straight into the camera saying, '...and it only does 20 miles to the gallon'. That's an 'up yours' to the environmental movement. It feels a bit dangerous, and it's really easy to do.

If you're told that the world is a scary place and you have to be responsible, having fun becomes all the more appealing.

Marcus We are more often exposed to advertising than we are to explicit works of art. And advertising is genius. The advertising that you don't think you've seen is even better. One of the things, perversely, that I think we should be hugely encouraged by is the success of bottled water. In the UK, there isn't anywhere you can't drink the tap water and yet we buy in plastic bottles something that's much more expensive than petrol. When petrol goes up by a penny it makes the news. They've convinced us to buy water from the Alps. You'd be very hard pressed to tell the difference between that and tap water. But we do it. We do it all the time. The marketing has worked on us. It's nuts that we buy bottled water. It's absolutely insane. Now if the environmental movement managed to harness that, and get people to do something which seems – from where we are now – crazy to a lot of people... there's a lot of encouragement to be taken from that.

Which found that in 2006 consumers in Britain spent £1.68bn on 2.275bn litres of bottled water, although consumption dropped in the following year: www.which.co.uk/environment-and-saving-energy/environment-and-greener-living/guides/switching-from-bottled-to-tap-water-/tap-vs-bottled-water. Researcher and activist Annie Leonard and colleagues followed up their highly successful 'story of stuff' animation (www.youtube.com/watch?v=9GorqroigqM) with an animated lecture on bottled water: http://storyofstuff.org/bottledwater/

So you take hope from human malleability and susceptibility to marketing? We can – even if we don't understand the reasons – be made to change our behaviour. New media is making a huge difference, isn't it? A blogger can have as much influence as a

columnist, a video can go viral in minutes and be seen by millions?

Joe If, ten years ago, you'd described a medium that was almost free to enter and allowed diverse public debate across the planet and that could cover environmental change, I would have said, 'Fantastic, just the job'. But one of the things we've seen is that sceptics or deniers or contrarians – I prefer the last term – have had an impact on mainstream media out of all proportion to their legitimate authority. That's a serious problem because there are influential chunks of populations, particularly young people, who are increasingly grazing for their content across these ranging plains.

The influence of climate contrarian bloggers on mainstream media can be explained in part in terms of their availability as a 'contrasting voice' at a time when climate change science and policy is viewed as orthodox and mainstream. The Ofcom judgement on Channel 4's transmission of the climate contrarian programme *The Great Global Warming Swindle* (2007) found that, despite the programme containing errors, the channel could escape censure for the broadcast under the terms of its public service obligation to challenge received views: climate change having now achieved the status of orthodoxy.

Here at Eden, you've made good use of the Web. Do you think it's a positive thing, or is there a risk that with so many messages people will graze between them and make up their own narratives?

Tim I think it identifies quite embarrassingly to what degree we are all sheep: the supposed opening up of democracy through the Web often indicates that people don't want to be different to everybody else so they will accept something without question. I'm hopeful for the Web but I don't think it's particularly exciting yet. What we're missing is that 'climate change' is a terrible title. 'Global warming' is a terrible title. We should stop talking about climate change in the way that we do. We've got to look for the battles we can win and then show how the quality of life is better for everybody. Don't even mention to the 'deniers' that it's an effort against them. I think we've got to be careful it doesn't become 'us' and 'them'. Let's just win one battle and once we've done that, ask, 'Do you all agree it's better not to see plastic bags blowing down the streets and turtles drowning?' Then we will do the next thing and the next – and suddenly a culture of the possible will emerge.

What would be the most successful avenues to explore in popular culture? An environmental story editor on *The Archers*? A character in *EastEnders* who is into recycling? Getting George Clooney to be spokesman for the planet?

The former agricultural story editor on the BBC Radio 4 serial *The Archers*, Graham Harvey, has written two books on environmental topics, *The Killing of the Countryside* (1998) and *The Carbon Fields* (2008), and the programme has touched on significant environmental issues, particularly through the experiences of Pat and Tony Archer, who converted to organic farming in the mid 80s, and their son Tom, arrested in 1999 for destroying an experimental field of GM rapeseed. In July 1999, a few months later, 28 Greenpeace activists were charged with criminal damage and theft when they 'decontaminated' a farm-scale trial of GM crops in Norfolk. They argued in court that they were acting to prevent pollen from the genetically-modified maize from polluting neighbouring organic crops and gardens and were found not guilty of theft; the jury failed to reach a judgement on the charge of criminal damage.

Joe I think it's a mistake to think that there is one place you will find this, and also that you can just insert a story line. The public has a great nose for authenticity. They've got to recognise themselves. The best [approach] would be to see everyday lives represented with the impact of carbon in the background of daily life. The trip to work: people spend 40 minutes in unpaid labour sitting in their steel boxes yet there is no opportunity to have a lovely bicycle, a dedicated cycle lane and showers at work, to cut gym fees, to have safe routes to school for kids. I'd like to see some of the great filmmakers, comics, drama writers, shine a light on the perversity of the last 60 years. We've won so little at such huge cost. Take it off the climate change topic. Of course it's there. It's in the background. Nestle it in our everyday concerns: food, sex, death and 'a nice day'.

Vicky I would encourage a culture of exploration, I think we are on a journey, an uncertain journey, and we need to delve into that and enjoy it and find out things. I think we expect answers, we expect solutions, unfairly often from scientists, but all of us now need to take a leap of faith and go forward and explore.

Marcus I would like to see more of anything that brings groups of people together in a way in which they are fully aware that they're surrounded by other people. The more we can develop a sense that the world is filled with other people who are worth meeting and whose space is just as important to them as mine is to me — that's an oblique way of bringing about environmental change: you can avoid the topic and people can have a really good time.

Artist Clare Patey has created a number of events that bring people together through food. *Feast*, a year-long project with Cathy Wren, involved children from a primary school taking responsibility for a nearby allotment and growing food for a community celebration, with other artists helping the children to create tablecloths, plates and performance. From this project, Patey developed *Feast on the Bridge*, an annual celebration that takes over Southwark Bridge in London, staging the whole life cycle of food from displays of beehives and breadmaking to stalls selling and giving away locally produced food, through to the collection of food waste for composting. Tables and chairs are laid down the length of the bridge, allowing some 40,000 strangers to sit down and eat over the course of the day. Patey describes it as a 'spectacular harvest supper that aims to reconnect urban people with the natural growing cycle'. See www.ashdendirectory.org.uk/features.

The jury for the 2004 Turner Prize praised Jeremy Deller's 'generosity of spirit across a succession of projects which engage with social and cultural contexts and celebrate the creativity of individuals'. As part of his exhibition at Tate Britain, members of Southwark Cyclists, a local cycle campaign, talked to hundreds of visitors in the gallery; on other days, a campaigner for bat conservation, a Quaker and historians of the labour movement and of the Empire Windrush also took up residence in the gallery.

Artsadmin's Two Degrees festival sits 'between art and activism, performance and protest', and offers 'a chance to be part of artist-led actions, tell your own revolutionary story, help eradicate invasive species, go on a mass bingo bike ride, ask an expert about the future or exchange personal and political views for a free haircut'. www.artsadmin.co.uk/projects/two-degrees.

Anatomy

The work of any artist thrives on the quality of the reactions it provokes. But to engage in any sustained and enlivening discussion about a work of art it is necessary to settle on a number of agreed terms. We know there is a difference between a horror movie and a romcom and these terms help us to avoid the sort of fruitless exchange in which oranges are compared to lemons. So when we discuss works about climate change, or works that have been influenced by climate change, we need to understand what kind of work each piece is. An exercise in agitprop might be said to be successful if it convinces you of its point of view, and even inspires you to take action, but a novel might need to be convincing in a totally different way. There's also a question about the terms themselves in relation to culture and climate change. Is there something about climate change which forces us to redefine the form as well as the content? Are the categories and genres in which we traditionally discuss works of art, TV programmes, novels and comedy, the appropriate ones for this emerging field?

This third discussion took place late on a November afternoon in one of the seminar rooms at the Open University in Camden, one of the OU's 13 regional centres in the UK. There was a small invited audience representing a range of groups with a lively interest in this field. On the panel were Bergit Arends, curator of contemporary arts at the Natural History Museum, Robert Butler, journalist and critic, Beth Derbyshire, artist, and Charlie Kronick, senior climate advisor for Greenpeace UK. The discussion was chaired by Quentin Cooper, presenter of BBC Radio 4's *Material World*.

Quentin began by suggesting that while classifying art forms can be tricky, it gets trickier still when those art forms are forms of cultural response. Not only do you have to stick a label on what the art form is, but there's also the business of what it is perceived as being a response to, and whether or not that is what the artist consciously or unconsciously intended.

Quentin Cooper: Can we start by trying to get a sense of the scale of this task? If we're looking right across the range of works about climate change, how far back does it go?

Charlie Kronick The relationship of artists to the environment goes back to cave paintings, because they're documenting the world that they're experiencing and the way that they experience it. More formally, the artists of the Romantic Movements, both in Europe and North America, and the Transcendentalists, love their environment and love talking about their relationship to it. I don't think it was so much about climate change as about the human relationship to the environment.

Does it affect the way you look at a work if you know it is about climate change?

Charlie My professional life is in the world of policy and the more technical and the drier end of the climate debate, but as a punter, if I'm looking at something created by an artist who is in some way moved or engaged with the change in climate, that's not the first thing I would think about. I'd be surprised if there were many people who did.

Bergit Arends I work at the Natural History Museum where there are many classification systems which co-exist. You can put things into categories, but then you can veer from one to the other and they are not mutually exclusive. We've come to think about art as something that we can classify into movements: 'surrealism', 'modernism' and so on. I don't think that holds true any more. Since the 1960s and 1970s, we've experienced what we call 'visual culture' and I want to use that as an idea for thinking about where

See *Cave of Forgotten Dreams* (dir. Werner Herzog, 2011) which explores the Chauvet caves in France. 'The time separating us from these artists is at least twelve times longer than the time separating us from the pre-Socratic philosophers. What makes their age astounding is the sensitivity of perception they reveal. The thrust of an animal's neck or the set of its mouth or the energy of its haunches were observed and recreated with a nervousness and control comparable to what we find in the works of a Fra Lippo Lippi, a Velazquez, or a Brancusi' (Berger, 2001:36).

—

Among those writers and artists who have come to be called the Romantics are 'a group of early nineteenth-century English poets who meditated deeply on the complex relations between humankind, nature and society: William Wordsworth, Samuel Taylor Coleridge, John Keats, Percy Bysshe Shelley, Lord Byron and John Clare'. They are the focus of Jonathan Bate's study, *The Song of the Earth* (2000).

—

The American writers identified as Transcendentalists include Ralph Waldo Emerson, Henry David Thoreau, Walt Whitman and Emily Dickinson.

—

Bergit Arends curated Mark Dion's 2007 'Systema Metropolis' exhibition at the Natural History Museum, in which Dion explored Linnaean classification systems, carrying out fieldwork at urban sites including Kingsnorth power station, the River Thames and the future Olympic site.

—

As an academic discipline, 'visual culture' draws on philosophy, cultural studies, critical theory and art history.

climate change fits in. We want to think about it much more as a cultural phenomenon rather than as an art historical term, to look into this activity thematically and think about what it is trying to achieve. Is it something activist? Is it something much more contemplative? It is something that can act as a metaphor? Is it a collaborative venture? I would look at this whole area with those questions in mind.

Beth Derbyshire I second that. Looking back to the 1970s, it's very interesting what happened: a lot of artists leaving the cities and galleries and going out to make work in the landscapes – and the work couldn't be sold. Some of that was about the gallery system at the time but some of it was also about man connecting with nature. When thinking about contemporary practice and responses, I don't know if you have 'climate change artists'. I think artists are working with the issues of their time and find a channel into this subject, their own way in. Climate change is so wide. Fifteen or sixteen years ago the topic had a very different reading, whereas now, because it's gone from the back of the newspaper to the front of the newspaper, there are all sort of ways into it, be it social, political... A number of artists work with trading commodities, like Amy Balkin for example. There are people who are looking at and working with environment in a totally different way – James Turrell's Roden crater, or Robert Smithson's Spiral Jetty or any of those iconic works of land art. Technology has exploded in the last eight years and the way that we interpret culture and ideas has changed radically. The impact that somebody has by climbing on the roof of Parliament may be best located on a blog or in the newspapers.

Robert Butler I would approach the 'climate change

Amy Balkin's *Public Smog* draws attention to the contested territory of the atmosphere. See www.publicsmog.org.

–

The Roden Crater is a project by James Turrell, constructed in an extinct volcanic cinder cone in the San Francisco Volcanic Field. Turrell purchased the land and started construction on the project in 1979. www.rodencrater.com.

–

The Spiral Jetty is an earthwork sculpture constructed from mud, basalt rocks and salt crystals by Robert Smithson in 1970, on the northeastern shore of the Great Salt Lake in Utah. www.robertsmithson.com.

–

Some rooftop protests around Westminster: on 20 March 2004, two Greenpeace activists protested on the roof of Parliament against the Iraq war; on 27 September 2005, a Fathers 4 Justice activist spent five hours on a roof; and on 27 February 2007, Plane Stupid campaigners protested on a roof against plans for a third runway at Heathrow. In October 2007, six Greenpeace climate change activists shut down the Kingsnorth coal-fired power station by occupying the smokestack. They argued in court that they were legally justified because they were trying to prevent climate change causing greater damage to property around the world. It was the first case in which preventing property damage caused by climate change had been used as part of a 'lawful excuse' defence and they were cleared of causing £30,000 of criminal damage. Documentary-maker Nick Broomfield made a film, *A Time Comes*, about the protest, and it was also the inspiration for a radio play, *The Kingsnorth Six*, by Julia Hollander.

artist' label in a different way, in that theatre already has an anatomy or taxonomy. It's got a lot of different genres and any playwright writing a play that touches on climate change will either be writing a comedy or a tragedy or a farce... and that affects the way they write the play and the meaning of what they're writing. Part of *The Contingency Plan* by Steve Waters (2009) is set in Westminster and Whitehall. At one point a drink is spilt over the Cabinet Minister's trousers and he has to take his trousers off and hang them on the radiator to dry. Well, now we're drawing on the Whitehall farce, the Brian Rix farce, of people walking around in their underpants. So the form of the play is also a part of the meaning of the play.

Does anything that is labelled as climate change art automatically lose some of its power because it will turn off those who are not interested in the subject?

Charlie I think it loses its power by becoming polemic. It can be great polemic but it loses its power to speak to a broader audience because it will categorise and polarise the response. It doesn't matter who labels it – it will still have the same effect. It has a divisive effect not just on the viewer but also on the makers. I've met artists who are profoundly moved by the impacts of climate change and are very, very worried about the implications, not just for their own family but for society at large, but the last thing they want is to be described as a climate change artist. They feel quite strongly that it undermines the integrity and authenticity of their own work. It's the composite societal impact that interests me: what brings the work into that wider conversation, to the wider political reaction you have as a viewer, as a voter, as a citizen, as a consumer.

First performed at the Bush Theatre, London in May 2009, and later broadcast on BBC Radio 3, *The Contingency Plan* was reviewed by Robert Butler for *Intelligent Life* magazine: http://moreintelligentlife.com/story/finally-good-play-about-climate-change. An interview with Waters is available at: www.ashdendirectory.org.uk/features.

—

Rachel Carson's *Silent Spring* (1962) is widely credited with inspiring the environmental movement. The book, first serialised in the *New Yorker*, documented the impact of DDT and other pesticides on wildlife and humans. Carson was a marine biologist by profession, and deeply influenced by literature and music. She had previously published books on sea-life and the shore, and later wrote on how to foster a sense of wonder in children. Critic Lawrence Buell described the tension between the literary and the polemic in her work: 'Any serious reader knows that the kind of art, the kind of criticism, which takes the form of a waywardly insinuating thought experiment can be more instructive than the overtly polemical kind which "has a palpable design on us", as John Keats dismissively put it. [...] But the two modes also need each other, even as they pull against each other. The reason that Rachel Carson's *Silent Spring* (1962) became one of those few works of environmental writing to have an immediate and significant impact on public policy, while several other well-researched contemporary books on the same topic did not, was that this was an author also capable of writing *A Sense of Wonder* (1964). And reciprocally: had celebration of nature's beauty been Carson's sole concern, her voice would have been lost to history' (Buell, 2005: vii).

Just as there will be scientists who choose to step beyond science and towards activism because they see climate change as a clear and present danger and want to alert people around the world, there may be artists who choose to go beyond being an artist and step towards activism. Do they risk the negative categorisation Charlie has just described?

Beth This issue is emotional, and it's human. You may as an artist or a scientist have a particular discipline, but this is something that affects us all and it's emotional because it questions our future.

Robert Climate and discussion about climate will percolate all through different artists' lives and the way they think. It's impossible not to think about it. So it will surface in many different ways, in very different genres. You don't want people to go and see a work of art and think because it's about climate or climate change it's part of a small group of works. You want to think that it's more like the impact of Darwin in the nineteenth century: it impacts right across the range.

Bergit If we take apart this idea of 'climate change art', it's looking at systems, because climate change is understanding systems; it's looking at our relationship to nature, and our observation of nature; it's looking at the impact of climate change, what caused it in the first place, at energy use and how we can address that. It's all of these. It's a term that needs so much unpacking that it's not very helpful to call something 'climate change art', because you want to look at all these different aspects. We can trace our cultural evolution, and look at our place in nature for a start. Where do we think we are within the natural environment? We

For example, science has been dramatised in many different genres, from performance art to Hollywood biopic. In *Science on Stage* Kirsten Shepherd Barr lists the 'wonderfully diverse' range of scientists who have peopled the stage since Brecht's *Galileo*, including Bohr, Heisenberg, Ernest Rutherford, Marie Curie, Lise Meitner, Ralph Alpher, Rosalind Franklin, Thomas Huxley, Tycho Brahe, Johann Kepler and Paul Dirac. http://ashdenizen.blogspot.com/2011/02/would-play-about-climate-scientists-be.html.

—

Gillian Beer shows how Darwin overturned fundamental cultural assumptions, and how novelists George Eliot and Thomas Hardy, among others, 'pursued and resisted the contradictory implications' of this shift, in her book *Darwin's Plots* (1983).

—

Key references that explore the heritage and consequences of the separation of nature and human culture include Clarence Glacken's *Traces on the Rhodian Shore* (1967), Bruno Latour's *We Have Never Been Modern* (1993), Stephen Toulmin's *Cosmopolis* (1992) and the volume of essays edited by William Cronon, *Uncommon Ground* (1997).

have considered ourselves outside of nature for quite some time, particularly since the eighteenth and nineteenth century. Are we beyond that? That's an important question.

Robert I agree that there has been a long tradition of work about nature, but I feel that this is a new moment. Any work about nature is dealing with it in a very different way. Seamus Heaney said that all poems now, any reference to nature, comes with a sense of nostalgia and loss because of this level of extinction that's going on. So it's more acute – it's a more acute moment.

We are all affected by climate change, but are we all equally effective in drawing attention to it? Are there some art forms, some areas of creativity, where there is a narrower range of response than others?

Robert In theatre, there have been probably only a dozen plays about climate change which, if you think of how long it's been an issue, is amazing. There have been plays that have responded quickly to the credit crunch, such as *Enron*. So there is something within theatre and the way it's currently presented that works against climate change as a subject.

Is it a limitation of the medium? You get films and poems about life underwater, but you don't get many plays because you can't put it on the stage so easily.

Robert Yes, but such profound philosophical questions are raised by climate change that I can't believe that theatre wouldn't want to address it. I don't mean in a campaigning way, I mean: how are people in an audience thinking about this, and thinking about what their values are?

In *Stepping Stones: Interviews with Seamus Heaney*, the poet tells Dennis O'Driscoll that 'environmental issues have to a large extent changed the mind of poetry. Again, it's a question of awareness, the horizon of consciousness within which poet and audience operate. There are those like Gary Snyder and Alice Oswald for whom these matters are an explicit concern, but at this stage nobody can have an uncomplicated Hopkinsian trust in the self-refreshing powers of nature' (O'Driscoll, 2008: 407). A little later in that interview he says, 'In terms of the distinction you made earlier between "environmental protest" and "ecological lament", I incline temperamentally to the latter. The poems are more like elegies for water and air than calls to action' (ibid: 411).

—

Enron by Lucy Prebble, first produced at the Royal Court Theatre, London, 2009. Other theatrical works on the theme in the same year include: *The Power of Yes: A Dramatist Seeks to Understand the Financial Crisis* by David Hare, and *A comic tries to understand the economic crisis* by Mark Thomas.

—

In the six months following this podcast several plays about climate change were staged in London alone, including *Greenland* at the National Theatre, collectively written by four playwrights (Moira Buffini, Matt Charman, Penelope Skinner and Jack Thorne) and *The Heretic* at the Royal Court Theatre (Richard Bean). These were reviewed by Robert Butler at: http://moreintelligentlife.com/content/arts/robert-butler/drama-climate-change.

Tony Kushner has said that as soon as you personalise climate change and begin to try to make it intimate you lose what it's all about. Was he saying that there are things you cannot do in dealing with climate change?

Charlie If you live in sub-Saharan Africa you've been in a generation-long drought – that is an experience of climate change. Equally, one of the privileged rich in the developed countries can make all kinds of consumer choices. That's also an experience – the way they buy a Prius or change their light bulbs – or don't. Whether those things are a conscious engagement with climate change or not, there is no doubt that you have a valid individual experience of climate change. I think it's perverse and unhelpful to say, 'Sorry, can't talk about it as a personal experience because that doesn't include the multitudes'. The multitudes are included whether Tony Kushner decides to do it or not. That's not a choice that they get to make.

It is that range of experience that makes it important for different people to engage with it. Climate change is a massively inert object. If you describe climate change as one big block, every time you push on it in one place it just doesn't move. That's why it's important that we experience it and engage with it across a range of activities and don't just say, 'Oh, it's got to be big' or 'It's got to be small.'

What if, rather than the artist bringing climate change to the wider public, the public brings climate change to the art work? I'm thinking, as a concrete example, of something like the exhibition 'The Ship: The Art of Climate Change' at the Natural History Museum. Ten years earlier or ten years later, I imagine there would have been a very different reaction.

Pulitzer prize-winning playwright Kushner reflects on the environmental legacy of his play *Angels in America* (2007), and discusses the artistic history and challenges of coping with climate change at: www.youtube.com/watch?v=75SCFOJnvWY.

—

'The Ship: The Art of Climate Change' (Cape Farewell/Natural History Museum, 2006) was co-curated by Bergit Arends and David Buckland of Cape Farewell, and included works by artists who joined the first three Cape Farewell expeditions to the Arctic. The exhibition toured nationally and internationally as 'Art and Climate Change' between 2006 and 2010. It was accompanied by the publication of *Burning Ice*, a book charting the experiences of the artists and the resulting work. See www.capefarewell.com/art/exhibitions/past-exhibitions/art-and-climate-change.html.

Bergit It was very interesting timing. The exhibition happened in summer 2006 and a few months before people didn't talk about climate change. But all of a sudden energy prices went shooting up and it was on everybody's mind. You would turn on the radio and you would hear somebody talking about climate change. It was a very interesting time for bringing climate change into popular thinking. The exhibition attracted over 100,000 visitors over 13 weeks. A museum becomes a place you trust, because there are a lot of people in this debate that you don't trust. As we've seen with the University of East Anglia debate, it's more convenient to shoot things down, to find flaws. People genuinely wanted to find out. We had the exhibition, and we printed a newspaper because it's got the sense of this is happening now – come and inform yourself. And people want to know.

It brought the museum to re-think what it stands for. It was the first time the Natural History Museum issued a statement on climate change; a second one followed on biodiversity. Institutions like that can play a really important role because they have scientific information, they've got a more neutral position within society. People will come to them for factual information and they have to take that role on because people are quite hungry for information. They want to know. And they also want to know what to do. I'm not sure if we could deliver that through an exhibition alone.

Charlie Just up the road from the Natural History Museum, at the Science Museum, there used to be a very interesting gallery about the importance of the change in climate and responding to it. This has recently been re-branded as 'The Shell Climate Change Gallery' which, funnily enough, is a lot less forthright about the need to respond to

The gallery is officially titled: *Atmosphere: exploring climate science.* The principal sponsors are Shell and Siemens. The museum received criticism from environmental NGOs for taking sponsorship from a fossil fuel company. Others noted that the museum had appeared to retreat 'into the science' and generally avoided engaging with policy and ethical discourses raised by climate science. See http://blog.platformlondon.org/content/shell-exhibition-science-museum-downplays-climate-urgency-targets-special-publics and www.sciencemuseum.org.uk.

climate change. Not, I would suggest, because of scientific uncertainties that have emerged as a result of Climategate, but because of the relationship between Shell as a sponsor and the cultural institution.

How do we begin to categorise some of the cultural responses to climate change? What are the most useful distinctions and classifications?

Robert As a critic, I think it's very important to judge a work on its own terms. So if you're judging a piece that's essentially a campaigning piece, that has very particular objectives compared to a piece that is more educational, is asking questions and is opening the subject up. I imagine with a campaigning piece you want to close the subject down, so that it boils down to a sentence that people can take home. Whereas, it should be possible to have a wonderful work that touches on climate change in various ways but you don't leave thinking you have to do anything. It was just absorbing. It was just a very fascinating topic. We need to go much further towards revealing the fascination of this topic. Much of the work, the ice cores and the tree rings, are things of beauty.

Charlie Take *An Inconvenient Truth*, which is obviously labelled as 'a film about climate change'. It's also a film about technological optimism. It's a film about the belief that the American way of doing politics, where you have a town meeting and you explain things to people and you overcome their deficit in understanding, will lead to a change in behaviour. It could be 'about' that, as much as about climate change. You can view any object through any lens that you choose. I think that what's really interesting about the negative response to Gore's

Climategate: journalistic shorthand to describe the hacking of over 1,000 emails and 3,000 other documents from the Climatic Research Group at the University of East Anglia, and the subsequent uproar on selective publication of the emails on websites in November 2009.

—

In 2010, in the wake of the Gulf of Mexico oil disaster, art activists staged a dramatic series of performances in cultural institutions to protest against oil companies like BP and Shell sponsoring gallery spaces like the Tate. 'Gushing from floral skirts, spilling elegantly from giant white eggs, jetting from paint tubes across the floor of the iconic Tate Turbine Hall, the flood of oily resistance that followed has generated a fierce debate in the art world around oil, ethics and sponsorship.' See Liberate Tate http://liberatetate.org/home.html, PLATFORM www.platformlondon.org and Art Not Oil www.artnotoil.org.uk.

—

The 'deficit model' of scientific communication is discussed further in Futures (page 84).

movie, is that it's not that people hate the idea that carbon dioxide has a greenhouse impact in the atmosphere. What they really hate is that it says, 'You are going to have to change'. They think it says, 'You are going to have to change the way you live'. You are going to have to accept a set of political values that you don't particularly like. That's what they don't like about that movie. They probably don't like Al Gore and they don't like his suits and they don't like his visual effects, but all those things are wrapped up together.

Is there more to be lost through classification than gained? You lose some of the effectiveness of art if you can say, 'Oh, that's a climate change artwork and therefore we don't need to engage with it'.

Beth It's a bit like the classic question, 'Is it art or isn't it?' That's not an interesting question. Is it interesting and is it valid? Activism is brilliant in many respects – and people who are both artists and activists often want to incite change or affect behaviour. Other people bring a different way to work. Going back to *An Inconvenient Truth*, I think it's also a film about conscience, and that's the thing that people find very difficult. One of the things that people find difficult when they think about climate change is that they are worried there is going to be a wagging finger. Where in fact, if you look at many people's work in this field, it's actually about revealing human insights and truths.

In 'Slow Activism: Homelands, Love and the Lightbulb', Wallace Heim describes the methodology of *Homeland*, a piece by the group PLATFORM which drew attention to 'three junctions in the provision of electric light', through a 'hybrid of activism, performance and conversation' (Heim 2003: 187). The article describes a 'loosely associated, international family of art works and actions known as social practice art, social sculpture, littoral art, dialogic art, new genre public art' and outlines the difficulties they pose for criticism from a single disciplinary perspective: in terms of the visual arts, they lack objects and take place over time; in terms of theatre, they are embedded in the everyday rather than fiction. They may have an underlying 'activist' drive to 'to transport participants towards a given or mutually decided change in thought or perception' but may work without obvious signs of 'activism' such as provocation, protest or demonstration.

Does a work of art about climate change need to have anything to do with reducing emissions?

Robert No, I don't think it does. I think that the world I am explaining to my children is completely different to the world I was born into, and climate change is the big reason it's different. So I want to hear the great minds, the great writers ponder this and write about it. That's what I want. I want a deeper engagement with the subject.

Charlie There are plenty of places you can go to find out how to reduce emissions. If it has some indirect non-linear outcome that leads to a reduction of emissions, hooray. But it would certainly not be on my report card as to whether it was a success or not.

You've got something here that's clearly evolving. It's gaining new limbs. It's losing organs. It's very difficult to put it into a particular pigeonhole. Shouldn't we be getting climate change art out there, rather than trying to work out what box to put it into?

Bergit Exactly. There's lots and lots of work out there – climate change art, or environmental art or ecological art – that deals specifically with climate change. This is definitely something we want to talk about, but there is a vast amount of cultural production that reflects our time and so will inevitably reflect climate change.

Beth What is clear is it can have a cumulative impact: the more good work that you see, the more effect you will have because you will reach more people. What's interesting is that so many different cultural practitioners are saying it in different ways, whether they are using newspapers, or boats, galleries or non-gallery work. I think that will lead to a collective voice and that should be encouraged.

Robert I broadly disagree with the question. I think the more you unpack it, the more you spread the goodies out on the table, the more you are able to differentiate and value what's special about different things. That's what I think of as 'anatomising', it's drawing out the wonder of each thing as a separate entity – instead of limiting, you liberate.

Charlie We are in a deeply utilitarian time. It's the age of austerity. Everything is going to be priced and put in the appropriate bin and if we can't afford it, it's going to go. That is going to make it very hard to respond to climate change. Art needs to get out there because of the impact it makes. It's in danger of being the thing that suffers first in the times we are in and I think it's important that it get out there, regardless of what you call it.

Futures

The way we imagine the future lies at the heart of any discussion about climate change. This fairly straightforward idea is complicated by the many aspects of climate change that threaten to undermine the way in which people in the 21st century have come to think about their futures. The implications of climate change pose some direct challenges, for instance, to current ideas of progress, economic growth, personal freedom and material self-improvement. It's understandable, therefore, that the science of climate change has been rejected by many who have another vision of the future. Our attitude to the future has been shaped by countless sources, from ancient philosophy and the great religions, through the Renaissance and the Enlightenment, to the latest movies and comics. But to what extent have we inherited from these diverse sources a way of thinking about the future that is particularly limiting in the light of these new circumstances?

This final discussion took place, two days after the Anatomy one, in the Governors' Room at the National Theatre Studio, where some of the great names in British theatre have met to discuss theatre programming, casting and touring. There was an invited audience which included artists, producers and writers. On the panel were Roger Harrabin, BBC environment analyst, Mike Hulme, author of *Why We Disagree About Climate Change*, Ruth Little, associate director of Cape Farewell, Oliver Morton, energy and environment editor for *The Economist*, and Carolyn Steel, architect and author of *Hungry City*. The discussion was chaired by Quentin Cooper, presenter of BBC Radio 4's *Material World*.

Quentin began by reminding the audience of the spectrum of ideas about the future. Climate scientists have outlined a range of possible scenarios, from one in which the odd degree rise in temperature brings forward the date that blossom comes out, to futures involving rising seas, floods, droughts, wars over resources and species extinction – including, some would argue, our own. While many people understandably prefer the 'early blossom' version of the future, non-governmental organisations and many others tend to focus on the more frightening end. The only thing that almost everyone can agree on is that – to slightly misquote Sam Cooke – an environmental change is gonna come.

Quentin Cooper: In what sense is climate science and policy 'cultural'?

Mike Hulme Undoubtedly, it is cultural. Climate science has oriented itself towards the future. It makes claims of greater or lesser veracity about how the future might unfold and because the future almost by definition is a contested place, it's a contested concept. First of all, it's going to become political – and, as we know, climate change has become deeply political – but because the future is a place that we all live in, in our imaginations, it also becomes cultural, in the sense that every single person on the planet has got a stake in the future and the way we think about the future, the way the future impacts back on us, how we connect to the future, our hopes and aspirations and fears. Climate change cannot escape from being both political and cultural. Whatever the scientists may think they're doing, they are actually invading that very contested place.

Oliver Morton At the moment, the most shared yet contested narrative about the future is a narrative about climatic change. It seems to be a narrative about stuff that happens independent of choice, and yet choice is a very important part of the politics that feed in to it. I was at a climate change meeting a couple of months ago and I realised that maps of the world with different places coloured in different levels of red are currently the most dominant graphic representation of the end of this century. We don't have lovely illustrations of skyscrapers and Zeppelins any more. What we have is maps of the world coloured different shades of red. That's a very important part of how we see the future.

Mike Hulme's argument is developed in his book *Why We Disagree About Climate Change: Understanding Controversy, Inaction and Opportunity* (2009). He writes in the preface to the book: 'The full story of climate change is the unfolding story of an idea and how this idea is changing the way we think, feel and act' (Hulme, 2009: xxviii).

—

The shades of red indicate global mean surface temperature rises, according to a range of scenarios.

Ruth Little I think science, like art, is a way of knowing and all ways of knowing are culturally mediated. I like the definition of culture that the anthropologist Clifford Geertz gave some years ago which is that 'man is an animal suspended in webs of significance he himself has spun'. I think that science and the way of knowing that science provides are as much culturally influenced and influencing as art itself, though communicated very differently – which is one of the big issues that we are now confronting in relation to climate change. Culture is a 'patterned practice'and we are all participants in it. Scientists, too, are members of society and have the same roles and responsibilities that the rest of us have in both engaging in and communicating potential futures for us.

'The concept of culture I espouse [...] is essentially a semiotic one. Believing, with Max Weber, that man is an animal suspended in webs of significance he himself has spun, I take culture to be those webs, and the analysis of it to be therefore not an experimental science in search of law but an interpretative one in search of meaning' (*The Interpretation of Cultures*, 1975: 5). See also note on Raymond Williams (page 52).

Carolyn Steel I agree. We can only see what we can see – there's a whole branch of philosophy that deals with this problem – but what's interesting about climate change is that this inconvenient truth is actually forcing us to re-think structures that have been quite comfortable, serving us quite well, for roughly the last two and a half centuries. We've inherited a series of structures – call it the Enlightenment – that has neatly split up the world and its problems into disciplines that can be approached with great professionalism. What we are facing now is to many people extremely scary. It's a level of complexity and engagement that requires that we re-think the very structures of society. And that's interesting.

Climate change is now a very familiar story, but how has that story evolved over 20 years of reporting?

Roger Harrabin I think the key element has been

fear. There have been many grand narratives that have interested the media and, through the media and NGOs, the general public and politicians. Before global warming came global cooling, over-population, fears of resource shortage... All these touch on people's imagination of their future and their children's future. I think it's natural that people would be concerned about these huge challenges facing us as a planet. It's been interesting to see how climate change has evolved because, unlike some of the other scares, it has gained traction rather than lost it over the years. Climate change has been underpinned by this great mass of climate science, although now we are starting to see a loss of confidence in some of that science because of Climategate, and because of the mistake on the IPCC report.

The other player in the cultural sphere is business. Oliver and I were at the CBI Climate Change Summit yesterday and one of the CBI business leaders, a CEO, said to me afterwards, 'We don't care what the sceptics are saying. As far as we're concerned this is a serious problem and we will keep driving it forward'. So, you have these different actors in society driving narratives forward at different paces, overlapping and feeding back into each other.

In the media, there is a definite problem of 'climate fatigue'. Over the past 20 years I've watched climate and the environment go up and down the news agenda as if on a roller coaster and that's why I'm so uncertain about where it goes now. I've just been doing a documentary in which we interviewed Michael Jacobs, who was Gordon Brown's advisor in the Copenhagen Climate Summit. And he says, 'Governments will really take climate change seriously, when climate change impacts upon them in a serious way'. Now when is that going to be? Are

Climategate: see footnote page 71.

—

This refers to an editorial error in the IPCC WG2 report that stated that the Himalayan glaciers could melt entirely by 2035. The date should have read 2350. This incident has led to debates about the fitness of the IPCC's current terms of reference and structure. The slow pace of the IPCC's response to these events also demonstrates the relatively light staffing of the secretariat and its modest press-handling resources.

—

The Confederation of British Industry (CBI) holds an annual conference on climate change. See http://climatechange.cbi.org.uk. In 2009, the Prince of Wales's Corporate Leaders' Group on Climate Change issued the 'Copenhagen Communiqué on Climate Change', supporting a robust deal at COP15. The communiqué secured the support of over 950 companies from the US, EU, Japan, Australia and Canada to Brazil, Russia, India, China and South Africa, ranging from the world's largest companies and best known brands, to small and medium-sized enterprises. www.copenhagencommunique.com.

we going to remain in a solar minimum? Are we going to go into a period of temporary cooling? What is going to happen with the level of warming with the planet? I have no idea. But by the time it impacts enough on nations for them to get worried – if the scientific account is correct – it is far, far too late. So his statement is not at all reassuring.

Oliver I think we need to look at that rather special sub-set of fear that we describe as the sublime. This idea that there is something frightening and yet there is something comforting about looking at the frightful thing because it's not actually proximate to you. You are saved by three floors of building above the flood. Mike's written about what's sometimes called 'climate porn', covers of newspapers saying 'Catastrophe! Six degrees warming!' and all that. There's a sense in which that fear is actually counter-productive because it's a way in which people externalise all sorts of other disaffections that they have with modernity, to say, 'Oh look at this terrible thing we are all doing' and to transfer it to an area, which they can condemn and worry about and have that frisson of fear, but not actually re-import into questions of, 'Well, does this actually affect the way I live my life?'

The concept of the sublime can be dated back to *On the Sublime*, a treatise written in the 1st century AD and attributed to Longinus. Eighteenth-century thinkers Kant and Burke developed the concept, linking the experience of the sublime to the difficulties we have in defining the object of such experience (due, for example, to the object's massive or unknowable scale – the depth of the sea or the height of a mountain). They defined the experience of the sublime as the taking of pleasure from the fear, wonderment, and dread arising from such difficulties.

–

The editor of Worldchanging.com, Alex Steffen, has blogged about the Renaissance concept of *terriblisma*, 'the strange, gratified awe one feels when beholding dreadful disasters and acts of God from afar', as a reference point for awed responses to climate change: 'We find ourselves riveted by strange occurrences and ominous portents – like giant squid growing to monstrous sizes in the warming greenhouse waters of our oceans, or the gigantic and ancient Larsen B Ice Shelf collapsing in Antarctica' (Steffen, 2003).

If you are talking about climate and climate futures you have to step beyond the facts. The facts will tell you where you are now but not the possibilities that they bring up for the future.

Oliver An important part of the debate is to extend that discussion of possibilities. Much of the discussion of climate change and activism around the issue is still about stopping it, about the limits that have been talked about since the 1990s by the

EU, and then increasingly by the rest of the world, of limiting climate change to two degrees. No one who is 'in the business' believes that this is a valid objective. And yet it is still seen – most famously at Copenhagen – as the campaigning point: that climate change is a future to avoid rather than a context for all the futures, good or ill, that we might go on to inhabit.

Mike It's a problem for scientists because as soon as you start talking about the future, you're leaving facts behind. Climate scientists have found this a very uncomfortable and difficult place to be. It's not surprising that different scientists position themselves in relation to these future narratives in different ways. So, we have someone like Jim Hansen who is a great scientist in the United States and has been working on this for 30 years or more. He will position himself in relation to future narratives in a very different place than someone like myself, who has also worked for most of my career on climate change. As soon as we start talking about climate change in the future, scientists inevitably have to bring other personal commitments or cultural commitments or sets of values and ethics into play and this has been difficult. It exposes a vulnerability for people who wish to critique or criticise science and scientists. It creates that vulnerability because people can say, 'Oh, but these guys, they're not talking about facts any more. They're talking about theories or predictions or projections.'

Oliver There's been a framing of this as an observational question, partly because of things like *An Inconvenient Truth*, whereas there's a very strong framing theory underlying much of it. When someone says, 'scientists can't say anything about the future

James Hansen's 1988 testimony to Congress began: 'Mr Chairman and committee members, thank you for the opportunity to present the results of my research on the greenhouse effect which has been carried out with my colleagues at the NASA Goddard Institute for Space Studies. I would like to draw three main conclusions. Number one, the earth is warmer in 1988 than at any time in the history of instrumental measurements. Number two, the global warming is now large enough that we can ascribe with a high degree of confidence a cause and effect relationship to the greenhouse effect. And number three, our computer climate simulations indicate that the greenhouse effect is already large enough to begin to effect the probability of extreme events such as summer heat waves [...] Altogether the evidence that the earth is warming by an amount which is too large to be a chance fluctuation and the similarity of the warming to that expected from the greenhouse effect represents a very strong case, in my opinion, that the greenhouse effect has been detected, and it is changing our climate now.' (Hansen, 1988). Hansen has been a lightning rod for climate contrarians as he remains one of the most high profile climate scientists to engage with politics and the media. He has recently produced a popular book on climate change (*Storms of My Grandchildren*, 2009) and testified at the Kingsnorth Six trial (see page 65).

with certainty'... well, scientists can say that next July will be warmer than next January; what you can't say is how warm it will be. There are many discussions, helped along by creationists, [which invoke the phrase] 'only a theory' – as though theory is essentially a second-rate and more culturally labile and weaker thing than having lots of dates about when tulips come into flower. In fact, theory is extraordinarily powerful.

Do we need to get away from this tendency to think about the future in utopian and dystopian terms?

Carolyn Yes. If you look at any set of human ideas it's always the one-liners that tend to get the most attention. It's much easier to generate newspaper headlines from big scary stories. We've inherited a polarised way of thinking from the Ancient Greeks: everything is presented as an either/or problem. I think what's required is a completely different way of thinking about problems, presenting them, arguing them.

Does this same polarity characterise artistic responses? In stage, film and TV drama, do we tend to present either shining, wonderful futures or deeply scary ones?

Ruth We are far more likely to present the deeply scary ones because they are so much more interesting. The problem, as Carolyn says, goes back to the dualism in western philosophy and in monotheistic religion. We've got good or bad and there's not much in between. Art has reflected that, but it has tended to err on the side of erring because that's where the drama is. A protagonist who is evil is more interesting. Milton knew that when he wrote *Paradise Lost*. Nobody is particularly interested in the character of Christ in *Paradise Lost*; Satan is so much more interesting and you'd much rather hang out with him in the wilderness.

I think we've got to accept that is a reality. The science fiction writer William Gibson argued that the future is 'already here. It's just not very evenly distributed'. We don't have to think too much about the future. We can look around us at the present and we can see this kind of dualism being played out in all of our imaginings of what we might become. Emerson argued that we can only see that which we are, and I think the business of framing is the quintessential problem: one of the reasons we cannot get a grasp on climate and climate change is that we frame our understanding of it. That's limiting. It's habitual. It's ideological. It's culturally constructed. If we've got to have conversations, artistic and imaginative conversations about the future, we've got to start by becoming conscious about what lens we are looking through. The big problem is that we don't realise we don't see those lenses any more.

Oliver I think we need to consider the historical context. Science fiction in the earlier part of the twentieth century has very optimistic traits, especially the American pulps. I think this is to some extent because it's a literature of colonising the future and is produced by first- and second-generation immigrants in New York; this is the science fiction that shapes our imaginations. But there is another future which is much more difficult, which is the future that comes with nuclear weapons. Nuclear weapons make the future either/or, dependent on something that literally happens in three minutes – the three minute warning. So, the future is either the radical preservation of the present – and we re-make the military industrial complex and change our notions of executive power in order to preserve the present – or unimaginably

Gibson said this in a discussion on US National Public Radio, 'The Science in Science Fiction' (30 November 1999).

–

In 1958, Rachel Carson described a state of mental paralysis which followed the explosion of the first atomic bombs: 'Some of the thoughts that came were so unattractive to me that I rejected them completely, for the old ideas die hard, especially when they are emotionally as well as intellectually dear to one. It was pleasant to believe, for example, that much of Nature was forever beyond the tampering reach of man, to suppose that however the physical environment might mold Life, that Life could never assume the power to change drastically – or even destroy – the physical world. These beliefs have been part of me for as long as I have thought about such things. To have them even vaguely threatened was so shocking that I shut my mind – refused to acknowledge what I couldn't help seeing' (Graham, 1970:13).

bad. That dominates our view of the future: we get utopias and catastrophes, the present preserved versus something horrid. That's the pre-history of the debate we are having now about how to forestall a future rather than how to navigate a future.

Roger I agree that the nuclear debate has helped frame the debate about climate change, but there are some more fundamental things here about the way we like to tell stories. In the broadcast media, and particularly in radio, we like to tell stories that take on ideas and topics through debate, and debate involves a natural, inevitable opposition. What we appear to have constructed in climate change is a bunch of people who say, 'I'm really worried about the future. I'm really worried about climate change'; a small group of people who say, 'I don't give a damn. It's not going to happen. Humans can't change the planet'; and quite a lot of people in the middle who say, 'Well actually, I don't know. I hear these competing voices and I don't know'. Now, there is another potential narrative which runs like this: 'There is massive consensus that humans have changed the planet already and will almost certainly change it some more. There is not a great deal of consensus about quite how future climate change will impact and what emission scenarios are tied to temperature outcomes and at the extreme end those scenarios are extremely scary and at the narrow end they are probably quite simple to cope with. We don't know which we are going to end up with'. That becomes more of a narrative of risk and risk avoidance and takes you into politics. Somehow we have failed to tell that [narrative]. It's not taken very long for me to tell it but over the years those of us in the media have failed properly to inform people about this issue because we're constantly pulled into

'Oh, we've got to have somebody saying something completely different'. Climategate was a real problem for the public consciousness. It seemed like something dodgy had gone on. Now I've looked very deeply into Climategate and I can't find any smoking gun at all. But I've also followed the enquiries into Climategate, and in my view they were all inadequate. So if you were looking on from the outside, from a suspicious viewpoint, you would be continuing to say, 'There is a scam. They are cheating us. The enquiries haven't looked into the issue properly' — because they haven't. It allows this continual erosion of a middle ground position. Politicians fall into the same trap. For the first time, at the 2010 CBI meeting, Chris Huhne talked about climate change in terms of risk and probability and insurance. Typically, until now, they've talked about climate change in terms of catastrophe. So, we may be starting to see a different narrative, but I'm not sure.

Secretary of State for Energy and Climate Change at the time of the 2010 CBI meeting.

—

A number of studies have shown the fallacy of the deficit model of communications. In relation to environmental issues, a classic paper is the discussion by Burgess, Harrison and Filius of the cultural politics of environmental citizenship (1998).

Mike Getting away from the metaphor of polarities or linear flows and replacing it with the metaphor of circularity would help at a number of levels. In science, there's been the linear model of how science should feed into policy; in science communication we have had the linear model of how science should be deposited in the minds of our citizens. But, if we displace those ways of thinking with notions of circularity then it allows different frames to emerge.

Is it more important to get change without understanding or understanding without change?

Mike Quite a lot of climate scientists still think in terms of the 'deficit model': if only we can have greater clarity, more access to the public, better science writers, then we will bring these recalcitrant and unruly people to book. We see plenty of evidence that climate scientists think in terms of this linear flow: 'we are the ones with the truth, the prophets who can see the future, and it's these people that we have to convince'. That is a deeply unhelpful way to bring science into public and political discussion. The metaphor of circularity, plurality, multiplicity, multi-vocality, is a much more engaging one. It gives us many more resources to work with creatively as a society or as a global collection of societies.

Ruth Art – in cultural forms of expression and self expression – can communicate an 'art of living' which contextualises our behaviour within the much broader frame of natural systems, living systems. Mike talks about developing new metaphors: I'm absolutely in agreement and think it has to happen at every level of cultural engagement. We desperately need to develop and articulate and communicate and find both physical and symbolic structures for these new metaphors. Circularity is a fantastic one because it's so closely tied in with living systems and seasons and cycles. I'm also interested in all the metaphors that chaos and complexity theory provide: the metaphors abound, from living beyond equilibrium, movement between order and disorder; everything that renders complex the nature of our engagement with the world and proves that we are contingent in everything we do, places us in an environment so that our actions aren't just single actions but they are 'actions towards' or 'actions in alliance with'. I think that we have framed the climate narrative as a negative one, of having to stop doing something. I'm much more interested in the possibilities of action in more complex contexts where single actions can generate unknowable and unpredictable outcomes, but they will generate outcomes, so action is worth taking.

Isn't there a tendency for us to imagine the future as much like the present, but subtly different, rather than considering that there might be a 'tipping point', which creates dramatic change so that we are suddenly facing a very different world?

Oliver Tipping points feel very frequently like a rhetorical stepping up, having lost the previous argument. You say, 'There will be warming.' Other people say, 'It won't be so bad.' And then you say, 'Ah-ha, there might be a tipping point.' Whether or not there is a tipping point, you run into this problem of how it feels as rhetoric.

The notion of 'the tipping point' was popularised by Malcolm Gladwell's 2000 book of that title. In terms of climate change, a key reference is a 2008 paper by Timothy Lenton *et al.*, reviewing nine potential 'Tipping Elements'. The UK organisation TippingPoint, which hosted the History discussion, was founded in 2005 with the objective of 'energising the creative response to climate change'. The organisation holds annual two-day events which bring climate experts together with artists. Working in partnership with the British Council, it has organised events across the world – in five continents by November 2011. TippingPoint has also awarded ten commissions for performance work, including the *Trashcatcher's Carnival*, *Third Ring Out* and *As The World Tipped*. See www.tippingpoint.org.uk.

Roger There is plenty of imaginative science that tries to look into the future. What we do know is that changes in technology, social organisation and political orientation are inevitably going to bring about dramatic and sometimes abrupt changes in the future – unless you think that the present social technological political state is solidified and will not change for the next hundred years. Here, some people describing climate science get it badly wrong. They seem to suggest that climate will unfold on a basically static planet and humanity. The science itself doesn't state that. If we are thinking about these non-linear or abrupt changes we should be thinking imaginatively about what may happen through social organisation, technology and politics. The collapse of the Soviet Union, the introduction of the Internet, the ability that we now have to manipulate our own genetic structures: those are dramatic, non-linear interventions in the world that might be just as important, I would suggest, as a hypothetical tipping point that may happen to the Amazon rainforest in 50 years time.

Ruth Don't they have a greater impact on us because things like genetic engineering threaten our notion of the integrity of our bodies? This is where I think that climate becomes real – it's the only kind of tipping point that really registers with most of us – when there's a threat from disaster, disease or death to the human body. That's when we start to acknowledge that we are endangered. At a TippingPoint conference in New York last year one of the artists said there is no point in having poster campaigns with polar bears sliding off melting icebergs, you have got to have poster campaigns in America of little blonde-haired, blue-eyed children drowning in Michigan. Until you reframe the narrative and say it's your body and your life that's endangered, nobody is going to care about what's happening because we are discounting distance and the future all the time. But we need both modes of expression if we're to communicate the complexity of human experience.

Can artistic responses help us prepare for the future, open our mind to the possibilities of mitigation, adaptation, etcetera?

Ruth I think they can and must, and that – for me – is the reason to be engaged as an artist. To be creatively engaged in the world is to engender and to support psychological potential for change and for evolution, for adaptation. Until knowledge is embodied, it's not received. It's not known. You can change people's behaviour through legislation: we've done it with seat belts and drink driving. But again that's about threatening the body – it's easy for us to understand why we are being harangued into doing these things. It's much harder to make a change in our lives for futures that we can't predict and we can't control, but it's easier if psychologically and emotionally we feel that there is a reason to do it. That's where art and culture can communicate in ways that science hasn't been allowed to in the past, because of scientific methodology and the ethical considerations around how scientists communicate their findings. Artists can do it on a human scale. I think art has a freedom of movement and manoeuvre that science can't have and shouldn't have if it's to remain something that can be trusted.

The Interdependence Day project brings together thinking about environmental and economic global changes. See www.atlas-id.org and the forthcoming *ATLAS of Interdependence* (Black Dog, 2012).

Do we need to have mental preparedness before we can have physical preparedness? Do artistic responses make us more willing to do new things or are they just as likely to paralyse us into doing nothing?

Carolyn I think what we need is mental openness – and that's almost the opposite of mental preparedness. What we need is the courage to say that we don't know, and this includes poor old scientists. I mean they get such a rough time. Why should one branch of humanity be expected to know all the answers? It's nonsense. We have to have a spirit of enquiry that we share. One of the key things is to open up to one another, discipline to discipline, branch of society to branch of society, nation to nation and say, 'We don't know, but we need each other.' So it's a radical reforming of interdependence, mutual dependence. It's very liberating to admit you don't know. Not enough people do it. I'm very interested in metaphor as a way of giving us sort of space to imagine new realities which don't necessarily have to be signed off by Thursday and costed by the next week.

Can cultural responses change the minds of politicians and the leaders of industry — perhaps more effectively than journalism, articles and data?

Roger Of course everybody is partly culturally determined. But you have to remember now that companies have set their own culture in which climate change, the low carbon economy, sustainability — for some of them — have become quite serious objectives and it would be difficult for them to begin unpicking that culture. There are various overlapping processes in society. Although I am not convinced about the way the climate change story will go, it's certainly far too early to write it off because there are so many powerful movers in society still pressing ahead with it. Just to put it into an international context, this great ferment of climate scepticism is a UK phenomenon — and partly also in the US, Australia and New Zealand — but for much of the rest of the world the understanding of the science has not changed.

All cultural responses, scales of intervention, direction have a part to play, but are some more effective than others? Are you going to change more minds through a blockbuster movie or through an art-house film?

Ruth There are three areas that I find very interesting. One of them is a cultural response to climate change which is largely elegiac. There is some absolutely beautiful and wonderful work coming out of that. The choreographer Siobhan Davies has made some fantastic pieces about imagining our future beyond the time when we have been able to continue to evolve and adapt, because

In Siobhan Davies' *Endangered Species* (2006) a small, semi-human figure dances gracefully inside a museum display case, her movements exaggerated by a costume of long bending rods that increase in number as her dance progresses. While at first they liberate her by extending the boundaries of her body, the many rods eventually restrict and finally extinguish her small life form. With its ever-increasing adornment and subsequent restrictions on expression, the dance points to how increased consumption alongside so-called technological 'advancement' is fast becoming more of a hindrance, rather than a help, to the development of our species. The figure put on show in a glass vitrine, re-emphasises the fragility of the dancing form and the need to preserve it. It is presented as a specimen, a rarity; a branch of the genus *Homo sapiens* that has either died out or is yet to evolve. The piece was made in collaboration with dancer Sarah Warsop, fashion designer Jonathan Saunders, film-maker Deborah May and artist and production designer Sam Collins. It was shown as part of the touring exhibition 'The Ship: Art and Climate Change' (see Anatomy page 69).

we've evolved to a standstill technologically, as it were, and materially. Then there is a fantastic campaign in New York by Joshua Allen Harris, a street artist, who has made a series of plastic bag polar bears which he puts over the gratings above the underground stations. When the trains go past, the air fills up the bag and the polar bear stands and comes to life and as the train disappears, it falls – in the most extraordinary, beautiful choreographed way and you watch the extinction of the animal on these New York subway gratings. It's a very powerful poetic, wordless way of envisaging the value of what we are losing. Secondly, there is the kind of work that is immersive, which I find fascinating. There are so many artists working in this field who have got inside natural processes. They've stopped standing back. They're often out of the galleries, though not always, they've communicated with scientists, sometimes collaborated very closely with them. They are making new forms and articulating new experiences of our world, our lives within the world. That art is very powerful because it shifts you psychologically and emotionally towards contextualised understanding of what it is to be human. Thirdly, where perhaps there is the most hope of all, is what I would call 'Leonardo' art, which is made by artists who are trans-disciplinary in their approach. They've said, 'No point in hanging around thinking of alternative futures. We've got to build them'. The example that I use is a company called Empower Playgrounds, which works in Ghana. They harness the kinetic energy of children to generate electricity from playground equipment to power their classrooms with computers and lights. It comes back to this idea of the 'art of living'. I like the idea of artists who willingly place themselves in the way of life in order to use the design competencies of art to engage

See www.empowerplaygrounds.org

with real problems and real possibilities.

Looking ten or a hundred years into the future, are you more optimistic or pessimistic in terms of our ability as a species to adapt and mitigate and postpone the effects of climate change? What part will culture play?

Carolyn The more I move in the circles where people are dealing with these issues, the more impressed I am by the creativity, forward thinking, inventiveness, adaptability and … I would use the word 'love'. People want to do the right thing and want to care. So, I have a lot of hope for humanity. I also have a healthy respect for the degree to which people can ignore problems – you know, the cancer patient smoking in bed – so I'm not optimistic about rapid change. I think that, as Ruth says, people will only respond when they can sense it being their problem and their body that's at risk.

Roger I'm not optimistic. I have noted that when I ask environmentalists, 'Are you optimistic?' they always reply 'I have to be optimistic'. In other words, they have to be optimistic otherwise they couldn't continue. I'm a journalist and I don't feel the need to be optimistic. I'm very pessimistic about global society's ability to deal with a challenge with this amount of uncertainty and complexity. There is also a scenario in which the world gets one and a half- or two-degree warming and actually a lot of places are greatly improved, because the warming will come in the northern latitudes mainly. But I don't want to take either an optimistic or pessimistic position. I don't feel it's helpful to throw your own human emotion into something of this nature.

Mike I am optimistic. I think I am religiously disposed

to be optimistic about the future, on condition that we get away from the hierarchy that C. P. Snow established in his book *The Two Cultures*, in which he elevated science and engineering and technology way above our cultural pursuits and our imaginative lives, valorising the sciences over the humanities in terms of their respective contribution to future human betterment. If we can allow our creative, imaginative and cultural potential to flourish at the same time as our scientific and technological creativity, then I'm optimistic.

Another reason for optimism is the sheer size of humanity on the planet. We have far more people alive today who are educated, exposed to a huge diversity of ideas and techniques and methods. If you think of that as creative potential, we are far more potentially creative than we have ever been on this planet before, and that suggests that we can find a whole variety and diversity of new ways of living on the planet. Climate change is the condition that we have to adjust to. It is simply the way the future is. It's not something to be stopped, to be controlled. The other thing about these more scary scenarios is the future never gets any closer. The future is always in the future. This is quite helpful — we do change our expectations of normality, sometimes incrementally, sometimes more rapidly. The future is there to draw us forward and I have an instinctive hope and optimism about humanity's future on the planet.

Ruth I think we will continue to face enormous stresses in population, poverty, energy use, environmental degradation and climate. Those things aren't going to go away; they're going to remain interlinked and mutually reinforcing. I don't have a lot of faith in the political process as it's currently established, on party political lines, and I don't have a lot of faith in corporations to stifle the desire for profit and think about resilience rather than growth. But I think that people are thinking about those things, ordinary human beings who haven't sacrificed themselves for the sake of a corporate profit motive. We are thinking about resilience in new ways that offer very promising opportunities to develop more diverse and sustainable societies.

See Thomas Homer-Dixon, *The Ingenuity Gap: Facing the Economic, Environmental, and Other Challenges of an Increasingly Complex and Unpredictable Future*, 2002.

Oliver I'm not sure about putting one's emotions into it. I'm not in my private life a particularly optimistic person. I am pretty optimistic about long-term outcomes in relation to climate, but I'm very pessimistic about the path by which they're reached. I think there are very bad ways to get there and better ways to get there. There is a useful concept introduced by the Canadian thinker Tad Homer-Dixon, 'the ingenuity gap'; there are solutions, but there is an ingenuity gap. I'm very worried about the degree to which we are going to be able to minimise suffering, specifically in the developing world, as we overcome that ingenuity gap. One of the things I have taken from this discussion is that climate change is a cultural focus, an 'imaginary' as sociologists say. I would like to see our imagination expand beyond climate, beyond how red or otherwise bits of the map are, and move instead into a more generalised concern with human development.

Contributors
Resources
Timeline

Contributors

Bergit Arends is curator of Contemporary Arts at the Natural History Museum where she has curated exhibitions including Lucy + Jorge Orta's 'Amazonia', Mark Dion's 'Systema Metropolis' and Cape Farewell's 'The Ship: The Art of Climate Change'. She also enables artists' residencies at the museum. From 1999 to 2004 she managed the science and art programme at the Wellcome Trust.

Marcus Brigstocke is a comedian, appearing regularly on TV and radio: *The Now Show* and *Just A Minute* (BBC Radio 4); hosting *The Late Edition* and *I've Never Seen Star Wars* (BBC Four) and appearing as team captain on *Argumental* (Dave). Marcus has played King Arthur in the UK tour of Monty Python's *Spamalot* and Perks the station porter in *The Railway Children* at Waterloo Station Theatre. His first book *God Collar* was published in June 2011. Marcus is an active campaigner for environmental awareness.

Robert Butler is a writer, specialising in two subjects – the arts and the environment. He was the theatre critic of the *Independent on Sunday* and writes regularly for theatre publications. He also writes the 'Going Green' column for *The Economist*'s *Intelligent Life* magazine. He is a Trustee of the Ashden Trust and co-editor of the *Ashden Directory*.

Nigel Clark is Senior Lecturer in Human Geography at the Open University. Nigel's research focuses on the variability and volatility of earth processes. He is interested in the way that people experiment with the various energies and materials they find around them and this interest in 'creativity' or 'generativity' links to a long-term engagement in the arts.

Quentin Cooper is best-known as the presenter of BBC Radio 4's *Material World*, the UK's most listened to science programme. He has written and presented many other TV and radio science series and is much in demand to present programmes, host conferences, give talks and run science communication workshops. He is also a film critic, author and former news, arts and youth programmes producer and presenter.

Siobhan Davies CBE is a choreographer and founder and director of Siobhan Davies Dance, one of the UK's leading independent dance companies. In 2006 Siobhan opened the RIBA award-winning Siobhan Davies Studios in south London. More than just a base for her company, the building represents Siobhan's vision of dance as a wellspring for ideas and creativity and is home to a wide range of collaborative and interdisciplinary projects.

Beth Derbyshire is an artist working primarily in film, through a collaborative and multi-disciplinary practice. Her subject matter embraces social issues and recent works have looked at our social and cultural relationship to environment. Beth's work has been exhibited by organisations including the Eden Project, the National Maritime Museum, the Gulf Art Fair, the Montreal Bienniale and Arco, Madrid.

Roger Harrabin is Environment Analyst for the BBC and a regular on radio programmes such as *Today* as well as commenting frequently on TV and online. Roger was founder and presenter of BBC Radio 4's environment magazine *Costing the Earth*. He has interviewed many leading figures including Margaret Thatcher, Tony Blair, Al Gore, John Kerry, Ban Ki-Moon, James Lovelock, Sonia Gandhi and Bjorn Lømborg. His two-part documentary, *An Uncertain Climate*, broadcast in 2010, examined media depictions of climate change.

Wallace Heim is a writer and researcher on performance and nature. Her doctorate was in philosophy, analysing methods for changing perceptions of nature – such as conversation, ethics and emotions. She is co-editor of the *Ashden Directory*, a director of the environmental research communication group PublicSpace, a trustee for PLATFORM, and a research fellow at the Social Sculpture Research Unit, Oxford Brookes University. She co-curated the conference Between Nature and co-edited *Nature Performed: Environment, Culture and Performance*.

Mike Hulme is Professor of Climate Change at the University of East Anglia and Founding Director of the Tyndall Centre for Climate Change Research. His work explores the idea of climate change using historical, cultural and scientific analyses, seeking to illuminate the numerous ways in which climate change is deployed in public and political discourse.

Charlie Kronick is Senior Climate Advisor at Greenpeace UK. He has worked in the fields of environment and development as an activist, campaigner, thinker and writer. He has focussed for the last decade on climate change-related issues; other work has included the international trade in hazardous waste, transport and road building, sustainable agriculture and the environmental and social impacts of corporate globalisation. He trained as a historian and historical bibliographer/ rare book librarian.

Ruth Little is a writer, dramaturg and Associate Director at Cape Farewell. Formerly, she lectured in English literature at the University of Sydney and was literary manager at the Royal Court Theatre. Ruth has developed a philosophy and methodology of 'metabolic' dramaturgy: the dramaturgy of non-linear living systems, which considers the biological, cognitive and social realms in the creation and interpretation of performance works, and attends to patterns, processes and emergent moments in dramatic structure and human experience.

Diana Liverman is Co-Director of Institute of the Environment, University of Arizona; Professor of Geography and Development, University of Arizona; Visiting Professor of Environmental Policy and Development, School of Geography and Environment, and Senior Research Fellow at the Environmental Change Institute, Oxford University. Her research interests include climate impacts, vulnerability and adaptation, and climate policy and mitigation especially in the developing world. She is a trustee of Cape Farewell and Julie's Bicycle.

Vicky Long is a producer and project manager. She is the producer for Cape Farewell of the Cape Farewell | Eden Project Collaboration and was Project Director of Cape Farewell from 2005 to 2009. Vicky has a background in theatre and scenography.

Eleanor Margolies is a writer and theatre-maker and wrote her doctoral thesis on objects and materials in performance. She is founder editor of *Puppet Notebook* magazine. Other publications include *Theatre Materials*, *The Day the Food Ran Out* and *Green Camberwell*.

Benjamin Morris is a writer and researcher whose work focuses on the relationship between culture, cultural heritage and the environment. His creative work appears widely in both the US and the UK, where he recently received his PhD in Archaeology from the University of Cambridge. Now an Affiliate Member of the OpenSpace Centre for Geographical and Environmental Research, he lives in New Orleans.

Oliver Morton is *The Economist*'s Energy and Environment Editor. He was previously the Chief News and Features Editor of *Nature*, the leading international scientific journal. He covers the energy business, climate science and policy, and other green issues. He is the author of *Eating the Sun: How Plants Power the Planet*, a study of the meanings and implications of photosynthesis, and *Mapping Mars: Science, Imagination and the Birth of a World*.

Kellie Payne is a PhD student at the Open University contributing to the Mediating Change project (supervised by Joe Smith and Robert Butler), investigating the relationship between culture and climate change. Previously, she worked in research communications in the area of agriculture and the environment and completed an MSc at the LSE.

Tim Smit KBE is Chief Executive and co-founder of the Eden Project. After a career in the music industry, Tim turned to the restoration of the Lost Gardens of Heligan in Cornwall and following that, the transformation of the exhausted Bodelva china clay pit into the Eden Project. Eden has pioneered creative and arts-led approaches to public education on the environment, believing that empathetic understanding can be more effective than knowledge transfer in triggering behavioural change.

Bradon Smith is a Research Associate in the Department of Geography at the Open University, working on climate change and culture. His background is in English literature, with doctoral research on popular science writing. He is currently Research Fellow on Climate Change for the Department for Culture, Media and Sport.

Joe Smith is Senior Lecturer in Environment at the Open University. His research interests lie in the history of environmental politics, public engagement and the media, and the politics of consumption – three areas which are not separate from each other but which instead inform each other. Joe's research practice has a strongly collaborative and interdisciplinary quality and an experimental edge.

Carolyn Steel is an architect, writer, lecturer, and director of Kilburn Nightingale Architects. She has taught at London Metropolitan University, at the London School of Economics where she was inaugural studio director of the Cities programme, and at Cambridge University, where she lectures on Food and the City. Her book, *Hungry City*, which won the Royal Society of Literature Jerwood Award for Non-Fiction, examines the relationship between food and the city.

Renata Tyszczuk is Senior Lecturer in Architecture at the University of Sheffield. Her current research and related art practice explores questions concerning global environmental change and provisionality in architectural thinking. She is a founding editor of *field:*, the online journal of architecture.

Resources

Selected organisations

Arcola Energy: information on the Arcola's vision of a 'carbon-neutral' theatre as well as practical steps for improving practice.
www.arcolaenergy.com

Artists Project Earth (APE) aims to create a better world by bringing the power of music and the arts to twenty-first century challenges. It supports effective projects and awareness raising initiatives to combat climate change and raises funds for natural disaster relief.
www.apeuk.org

Ashden Directory is a UK directory of companies – with articles and the Ashdenizen blog – bringing together environment and performing arts.
www.ashdendirectory.org.uk

Atlas of Interdependence gathers essays, images and broadcasts generated by the Interdependence Day project, a partnership between Open University Geography, University of Sheffield Architecture and the New Economics Foundation that sought fresh responses to global environmental change and globalization.
www.atlas-id.org

Cape Farewell brings artists, scientists and communicators together to stimulate the production of art founded in scientific research.
www.capefarewell.com

Carbon Arts is a non-profit arts organisation that is working to facilitate an increased role for artists in generating awareness and action on climate change.
www.carbonarts.org

Centre for Sustainable Practice in the Arts is a centre within CAL ARTS.
www.sustainable practice.org

Creative Climate is a diary project recording how people understand and respond to environmental change.
www.creativeclimate.org

Goethe Institute Culture and Climate Change website: articles, international bloggers and a listing of 40 recommended films.
www.goethe.de/climate

Green Museum offers archive notes and news on environment-related art works, intended as 'a giant collaborative art-making tool'.
www.greenmuseum.org

Greening Theatre: a UK group for theatre professionals interested in tackling climate change, in response to the Green Theatre Plan.
www.greeningtheatres.com

Green Theater Initiative: a US-based website to inspire theatres to incorporate environmental concerns in their planning and operations.
www.greentheaters.org

Julie's Bicycle: a not-for-profit company helping the performing arts (especially the music and theatre industries) to cut greenhouse gas emissions and create a low carbon creative future.
www.juliesbicycle.com

Landscape and Environment: a programme of events, research and publication sponsored by the UK Arts and Humanities Research Council. Many of the projects concern the contributions of the arts and humanities to exploring climate change.
www.landscape.ac.uk

Mediating Change: Environment, Culture and Politics is a research group based in the OpenSpace Research Centre with participants from Open University Geography, University of Sheffield Architecture and the Ashden Trust.
www8.open.ac.uk/researchcentres/osrc/research/collaboratories/mediating-change-environment-culture-and-politics

PLATFORM works across disciplines for social and ecological justice and 'combines the transformative power of art with the tangible goals of campaigning, the rigour of in-depth research with the vision to promote alternative futures'.

www.platformlondon.org

Sculpt the Future aims to promote positive environmental change towards global sustainability by supporting creative, innovative and sustainable action. http://sculptthefuturefoundation.org

Tara Expeditions conducts scientific and communications work from a sailing boat to win media attention for climate change, particularly in France. http://arctic.taraexpeditions.org

TippingPoint aims to 'harness the power of the imagination to help stabilise the climate' with dialogue between artists, scientists and others close to the heart of the issue. www.tippingpoint.org.uk

Selected viewing

Koyaanisqatsi (1983) dir. Godfrey Reggio; music by Philip Glass.

The Day After Tomorrow (2004) dir. Roland Emmerich.

Les Glaneurs et La Glaneuse (2004) dir. Agnès Varda.

An Inconvenient Truth (2006) dir. Davis Guggenheim; featuring Al Gore.

Art From a Changing Arctic (2006) dir. David Hinton

The 11th Hour (2007) dir. Leila Connors and Nadia Connors; narrated by Leonardo DiCaprio, with contributions from over 50 politicians, scientists, and environmental activists.

Everything's Cool (2007) dir. Daniel Gold, Judith Helfand; with Bill McKibben.

Home (2009) dir. Yann Arthus-Bertrand.

Avatar (2009) dir. James Cameron.

The Age of Stupid (2009) dir. Franny Armstrong; featuring Pete Postlethwaite.

Into Eternity (2009) dir. Michael Madsen www.intoeternitythemovie.com

Climate of Change (2010) dir. Brian Hill, screenplay Simon Armitage, narrator Tilda Swinton.

The Story of Stuff www.storyofstuff.com

The Most Terrifying Video You'll Ever See 2 www.youtube.com/watch?v=AE6Kdo1AQmY

Selected reading

Jonathan Bate (2000) *The Song of the Earth*, London: Picador.

David Buckland *et al.* (2006) *Burning Ice: Art & Climate Change* London: Cape Farewell.

David Buckland and Chris Wainwright (eds) (2010) *Unfold: A Cultural Response to Climate Change*, New York and Vienna: Springer.

Nigel Clark (2010) *Inhuman Nature: Sociable Life on a Dynamic Planet*, London: Sage.

William Cronon (1997) *Uncommon Ground: Rethinking the Human Place in Nature*, New York: W. W. Norton.

Field: Ecology (2011) Available at: www.field-journal.org/index.php?page=issue-4

Clarence Glacken (1967) *Traces on the Rhodian Shore: Nature and Culture in Western Thought from Ancient Times to the End of the Eighteenth Century*, Berkeley and Los Angeles: University of California Press.

Mike Hulme (2009) *Why We Disagree About Climate Change: Understanding Controversy, Inaction and Opportunity*, Cambridge: Cambridge University Press.

Elizabeth Kolbert (2006) *Field Notes from a Catastrophe: Man, Nature, and Climate Change*, London: Bloomsbury.

Richard Mabey (2005) *Nature Cure*, London: Pimlico.

Bill McKibben (2011) *The Global Warming Reader*, New York: OR Books.

Lucy Neal and Hilary Jennings (2010) *Sustainable Ability: Mapping the Response to Resource Scarcity and Climate Change, Best Practice and Barriers to Greater Response*. Available from www.sustainableability.com/report.

Stephen Peake and Joe Smith (2009) *Climate Change: From Science to Sustainability*, Oxford: Oxford University Press.

Jonathan Porritt, Francesco Manacorda and T.J. Demos (2009) *Radical Nature: Art and Architecture for a Changing Planet*, 1969-2009, Cologne: Walther König.

Alex Steffen (2011) *Worldchanging: A User's Guide for the 21st Century*, New York: Harry N. Abrams.

Kathryn Yusoff, ed. (2008) *Bipolar*, London: Arts Catalyst.

Works cited in this volume

Bate, Jonathan (2000) *The Song of the Earth*, London: Picador.

Beer, Gillian (1983) *Darwin's Plots: Evolutionary Narrative in Darwin, George Eliot and Nineteenth-Century Fiction*, Cambridge: Cambridge University Press.

Benjamin, Walter (1999) 'The Lisbon Earthquake', in *Selected Writings vol. 2*. Boston: Belknap, Harvard University Press.

Berger, John (2001) *The Shape of a Pocket*, London: Bloomsbury.

Boyce, Tammy and Justin Lewis (2009) *Climate Change and the Media*, New York: Peter Lang.

Boykoff, Maxwell T. (2011) *Who Speaks for the Climate?: Making Sense of Media Reporting on Climate Change*, Cambridge: Cambridge University Press.

Buckland, David *et al.* (2006) *Burning Ice: Art & Climate Change*, London: Cape Farewell.

Buell, Lawrence (2005) *The Future of Environmental Criticism*, Oxford: Blackwell.

Burgess, Jacquelin (1990) 'The Production and Consumption of Environmental Meanings in the Mass Media: A Research Agenda for the 1990s', in *Transactions of the Institute of British Geographers*, 15 (2), pp. 139-161.

Burgess, Jacquelin, C. M. Harrison and P. Filius (1998) 'Environmental Communication And The Cultural Politics Of Environmental Citizenship', in *Environment and Planning A*, 30(8), pp. 1445-1460.

Burroughs, W. J. (2005) *Climate Change in Prehistory: The End of the Reign of Chaos*, Cambridge: Cambridge University Press.

Butler, Robert, 'A Greener National Theatre' <www.nationaltheatre.org.uk/57970/behind-the-scenes/a-greener-national-theatre.html> [Accessed May 18, 2011].

Byers, Michael (2005) 'On Thinning Ice' *London Review of Books* Vol. 27 No. 1, pp. 9-11. Available at <www.lrb.co.uk/v27/n01/michael-byers/on-thinning-ice>

Calvin, William H. (2002) *A Brain for all Seasons: Human Evolution and Abrupt Climate Change*, Chicago: University of Chicago Press.

Carson, Rachel (1962) *Silent Spring*, London: Hamish Hamilton.

Clark, Kenneth (1973) *The Romantic Rebellion: Romantic Versus Classic Art*, London: John Murray.

Clark, Nigel (2002) 'Feral Ecologies: Performing Life on the Colonial Periphery' in *Nature Performed: Environment, Culture and Performance*, Oxford: Blackwell.

— (2010) *Inhuman Nature: Sociable Life on a Dynamic Planet*, London: Sage.

Cohen, Stanley (2000) *States of Denial: Knowing About Atrocities and Suffering*, Cambridge: Polity Press.

Corporate Leaders Group (2009) *The Copenhagen Communiqué on Climate Change*, <www.copenhagencommunique.com/the-communique> [Accessed 17 June 2011].

Daniels, Stephen and Georgina H. Endfield (2009) 'Narratives of climate change: introduction', in *Journal of Historical Geography*, 35(2), pp. 215-222.

Flannery, Tim (2011) 'We will form a global community with a set of shared beliefs' www.guardian.co.uk/profile/tim-flannery [Accessed 20 June 2011]

Funtowicz, Silvio and Jerome Ravetz (2003) 'Post-Normal Science'. Available at: <www.ecoeco.org/education_encyclopedia.php>.

Geertz, Clifford (1973) *The Interpretation of Cultures*, London: Hutchinson.

Gladwell, Malcolm (2000) *The Tipping Point: How Little Things Can Make a Big Difference*, London: Little, Brown.

Gorky, Maxim ([1901] 2007) *Philistines*, in a new version by Andrew Upton, London: Faber.

Graham, Frank (1970) *Since Silent Spring*, Boston: Houghton Mifflin.

Hansen, Anders (2009) *Environment, Media and Communication*, Abingdon: Taylor & Francis.

Hansen, James (1988) 'The Greenhouse Effect: Impacts on Current Global Temperature and Regional Heat Waves' Report to United States Senate Committee on Energy and Natural Resources. Available from <www.guardian.co.uk/

environment/2008/jun/23/climatechange.carbonemissions?INTCMP=SRCH> [Accessed 20 June 2011].

Heim, Wallace (2003) 'Slow Activism: Homelands, Love and the Lightbulb' in *Nature Performed: Environment, Culture and Performance*, Oxford: Blackwell.

Henry, Kevin (2007) 'Parallel Universes: Making Do and Getting By + Thoughtless Acts (Mapping the Quotidian from Two Perspectives)' in Core 77 <www.core77.com/reactor/03.07_parallel.asp> [Accessed December 12, 2007].

Homer-Dixon, Thomas (2002) *The Ingenuity Gap: Facing the Economic, Environmental, and Other Challenges of an Increasingly Complex and Unpredictable Future*, New York: Random House.

Hulme, Mike (2009) *Why We Disagree About Climate Change: Understanding Controversy, Inaction and Opportunity*, Cambridge: Cambridge University Press.

Humphreys, David (2011) 'Smoke and Mirrors: Some Reflections on the Science and Politics of Geoengineering', in *The Journal of Environment & Development*, 20 (2): 99-120.

Ingram, John, Polly Ericksen and Diana Liverman (eds.) (2010) *Food Security and Global Environmental Change*, London: Earthscan.

Lamb, Hubert (1995) *Climate, History and the Modern World*, London and New York: Routledge.

Leiserowitz, Anthony (2004) 'Before and After The Day After Tomorrow', in *Environment*, 46(9), pp. 22-37.

Lenton, Tim *et al.* (2008) 'Tipping elements in the Earth's climate system' in *Proceedings of the National Academy of Sciences*, 105(6), pp. 1786-1793.

Liverman, Diana and Simon Billett (2010) 'Copenhagen and the Governance of Adaptation', in *Environment* May/Jun 2010. 52(3) pp. 28-36.

Liverman, Diana (n.d.) *WRT: Informing Adaptation* <www.worldresourcesreport.org/responses/informing-adaptation> [Accessed 17 June 2011]

McEwan, Ian (2010) *Solar*, London: Cape.

Meadows, Donella H. (1972) *The Limits to Growth: A Report to the Club of Rome*, New York: Universe Books.

New, Mark, Diana Liverman and Kevin Anderson (2009) 'Mind the Gap: Policymakers Must Aim to Avoid a 2 °C Temperature Rise, But Plan to Adapt to 4 °C', in *Nature (Reports Climate Change)*, pp. 143-144.

O'Driscoll, Dennis (2008) *Stepping Stones: Interviews with Seamus Heaney*, London: Faber and Faber.

O'Riordan, Tim *et al.* (2011) *Metaphors of Tipping Points*, Unpublished paper presented at the Tipping Points workshop, Royal Society Kavli Centre, UK.

Populus (2010) *BBC Climate Change Poll*, <http://news.bbc.co.uk/nol/shared/bsp/hi/pdfs/05_02_10climatechange.pdf> [Accessed 19 June 2011]

Richardson, Katherine, Will Steffen and Diana Liverman (2011) *Climate Change: Global Risks, Challenges and Decisions*, Cambridge: Cambridge University Press.

Rose, Chris (2010) 'Climate Change Campaigns: Keep Calm But Don't Carry On'. Available at <www.campaignstrategy.org/> [Accessed June 17 2011]

Serres, Michel (1995) *The Natural Contract*, Ann Arbor: University of Michigan Press.

Simms, Andrew (2008) 'We have only 100 months to avoid irreversible environmental disaster' <www.guardian.co.uk/environment/2008/aug/01/climatechange.carbonemissions> [Accessed May 24, 2011].

Simms, Andrew and Joe Smith, eds. (2008) *Do Good Lives Have to Cost the Earth?* London: Constable.

Smith, Joe, Nigel Clark and Kathryn Yusoff (2007) 'Interdependence', in *Geography Compass* vol. 1, no. 3, pp. 340-359.

Smith, Joe ed. (2000) *The Daily Globe: Environmental Change, the Public and the Media*, London: Earthscan.

Snow, C. P. (1993) *The Two Cultures*, ed. Stefan Collini, Cambridge: Cambridge University Press.

Solnit, Rebecca (2011) 'The Butterfly and the Boiling Point: Charting the Wild Winds of Change in 2011'. Available at www.TomDispatch.com.

Spence, Alexa *et al*, 'Public Perceptions of Climate Change and Energy Futures in Britain: Summary Findings of a Survey Conducted in January-March

2010'. Available at <www.ipsos-mori.com/Assets/Docs/Polls/climate-change-public-perceptions-of-climate-change.pdf> [Accessed June 20 2011]

Steffen, Alex (2003) 'Terriblisma'. Available at <www.worldchanging.com/archives/000089.html> [Accessed April 1, 2011].

Stern Review (2006) *The Economics of Climate Change*, Cambridge: Cambridge University Press.

Thornes, John (1979) 'Constable's Clouds', in *The Burlington Magazine*, vol. 121, 920 pp. 697 -699 + 701-704.

– (1999) *John Constable's Skies: A Fusion of Art and Science*, London: Continuum.

Toulmin, Stephen E. (1992) *Cosmopolis: The Hidden Agenda of Modernity*, Chicago: University of Chicago Press.

Weart, Spencer (2008) *The Discovery of Global Warming*, Cambridge MA: Harvard University Press.

Williams, Raymond (1976) *Keywords: A Vocabulary of Culture and Society*, London: Croom Helm.

Wittgenstein, Ludwig (1980) *Culture and Value*, G. H. von Wright, ed., London: Blackwell.

Timeline

1200s
Collapse of the Ancient Pueblo or Anasazi civilisation in Southwest USA attributed to drought and deforestation.

1608
The first recorded frost fair on the Thames; the Thames froze for weeks at a time in many winters between the mid-14th and 19th century, in the period known as the 'Little Ice Age'.

1661
In *Fumifugium, or, The Inconveniencie of the Aer and Smoak of London Dissipated*, John Evelyn proposes remedies for London's air pollution including public parks and the planting of flowers and fragrant trees. In *Sylva, or a Discourse of Forest Trees* (1664) he encourages landowners to plant trees to supply the navy.

1712
Invention of the atmospheric steam engine by Thomas Newcomen (later improved upon by James Watt) leading to the Industrial Revolution and increased emission of greenhouse gases due to burning coal, steam railways and land clearance.

1755
Massive earthquake destroys Lisbon and shakes belief in benevolent nature among Enlightenment thinkers.

1776
The History of the Decline and Fall of the Roman Empire by Edward Gibbon is 'the first systematic attempt to relate climatic factors to the declining fortunes of a major civilisation' (Hulme, 2009: 28).

1783
A volcanic eruption in Iceland causes a 'year without a summer'.

1800–1870
Level of carbon dioxide gas (CO_2) in the atmosphere, as later measured in ancient ice, is about 290 ppm (parts per million). Mean global temperature (1850-1870) is about 13.6 °C.

1803
Luke Howard publishes *Essay on the Modification of Clouds*. His classification and naming of clouds influences Shelley, Constable and Ruskin.

1836
'Nature' by Ralph Waldo Emerson, an essay that influences Henry David Thoreau, Emily Dickinson and other writers later described as Transcendentalists.

1854
Walden; Or, Life in the Woods by Henry David Thoreau records his experience of simple living and self-sufficiency.

1864
Man and Nature by George Perkins Marsh – an early but influential work documenting human effects on the environment.

1889
Writer and naturalist John Muir begins the campaign to save the Yosemite region in California from exploitation. His articles in *Century Magazine* lead to a bill in Congress to expand federal protection, and ultimately to the creation of the National Park Service in 1916.

1890–1920
Composers including Delius, Vaughan Williams and Percy Grainger contribute to the English Folk Revival, celebrating an endangered rural life. Its detractors called it 'cowpat music'.

1896
The National Trust is founded by Octavia Hill, Robert Hunter and Hardwicke Rawnsley to conserve threatened coastline, countryside and buildings. They are supported by John Ruskin, artist, writer and advocate of conservation measures including town and country planning, green belts and smokeless zones.

–

Arrhenius publishes the first calculation of global warming from human emissions of CO_2.

1915
The Origin of Continents and Oceans by Alfred

Wegener introduces tectonic plate theory, arguing that 300 million years ago the continents formed a single landmass.

1920–1925
Opening of Texas and Persian Gulf oil fields.

1925
The Professor's House by Willa Cather describes the abandoned cities of the Anasazi.

1926
Russian scientist Vladimir Vernadsky publishes theory of the integration of the biosphere, or living matter, and the earth's geological processes.

1936
Noah and the Waters by Cecil Day Lewis imagines the inundation of contemporary Britain.

1946
Founding of the Soil Association in the UK.

1951
The Day of the Triffids by John Wyndham: bioengineered plants – mobile, carnivorous, poisonous – escape into the wild.

1953
The Man Who Planted Trees, by Jean Giono, a fable of the reforestation of a French valley.

–

Smog - a mixture of smoke and fog - is blamed for the deaths of 1,000 Londoners, forcing the British Parliament to pass the Clean Air Act.

–

In January, a storm tide devastates the coasts of East Anglia and the Netherlands, killing more than 2,000 people.

1955
Patrick Hadley's cantata *Fen and Flood*, arranged for four-part chorus by Vaughan Williams, a history of East Anglia and its relationship with the sea, responds to the 1953 floods.

1956
The Death of Grass by John Christopher: a virus that wipes out grass and crops decimates Asia, causing mass starvation and riots, and eventually hits Britain.

–

Mysterious illnesses in the small town of Minimata in Japan are found to be caused by mercury poisoning from industrial pollution. Attempts at a cover-up by business and government create the world's first environmental scandal. David Holman's 1972 play *Drink the Mercury* portrays the aftermath of the poisoning.

1957
The International Geophysical Year helps to establish a global scientific community concerned with exploring planetary processes. This lays the ground for later studies that point to human-induced climate change.

1960
Charles David Keeling accurately measures CO_2 in the earth's atmosphere and detects an annual rise. The level is 315 ppm. Mean global temperature (five-year average) is 13.9 °C.

–

In 'Dome over Manhattan' Richard Buckminster Fuller and Shoji Sadao sketch a huge dome covering a large proportion of Manhattan island to create a climatically self-sufficient city.

1962
Silent Spring by biologist Rachel Carson popularises understanding of the impact of pesticides on wildlife and humans. The grassroots environmental movement the book inspires leads to the creation of the Environmental Protection Agency in the United States.

–

Peter and Eileen Caddy and Dorothy Maclean found the Findhorn community in Scotland.

–

The Drowned World by JG Ballard imagines a flooded London after ice caps melt and sea levels rise.

1963
US Congress passes the first Clean Air Act.

1966
Founding of *Resurgence* magazine, which explores ecology, spirituality and the arts. Early contributors include E.F. Schumacher, Leopold Kohr and John Seymour.

1967
The Torrey Canyon oil tanker breaks open off the coast of Cornwall - the first major oil spill.

1968
The Whole Earth Catalog provides a handbook for self-sufficiency, listing equipment, tools and machinery, alongside articles on topics including organic farming, resource depletion, solar power, recycling and wind energy.

–

Elisabeth Beresford publishes the first Wombles book, which in 1973 becomes a popular UK children's TV show. The Wombles' theme song includes the lines 'Making good use of the things that we find,/Things that the everyday folks leave behind' and their motto is 'Make Good Use of Bad Rubbish'.

1969

Friends of the Earth founded in the USA.

—

Operating Manual for Spaceship Earth by Buckminster Fuller popularises the phrase 'Spaceship Earth' to describe the planet's finite resources.

1970

First Earth Day.

—

The Ecologist magazine founded.

1971

Greenpeace founded in Canada.

—

The Lorax by Dr Seuss describes what happens when a forest of Truffula trees is chopped down.

—

Design for the Real World by Victor Papanek draws attention to the damage caused by corporations and consumer culture and provokes debate about the ethical responsibilities of design practice.

—

Not Not Not Not Not Enough Oxygen, a play by Caryl Churchill set in London 2010.

—

Diet for a Small Planet by Frances Moore Lappé exposes the waste in US grain-fed meat production.

—

What's Going On, a concept album with a strong environmental theme, by Marvin Gaye.

1972

The United Nations Conference on the Human Environment (the Stockholm Conference) produces the first document in international law to recognise the right to a healthy environment. The United Nations Environment Programme is formed. Many in developing countries accused environmentalists in the developed world of 'pulling up the ladder behind them'.

—

Droughts in Africa, Ukraine and India cause world food crisis.

—

The American meteorologist Edward Lorenz presents a paper, 'Predictability: Does the Flap of a Butterfly's Wings in Brazil Set Off a Tornado in Texas?' pointing out the chaotic nature of climate systems and the possibility of sudden shifts.

—

Memoirs of a Survivor by Doris Lessing, a dystopian novel about the breakdown of society.

—

Mike Reynolds builds his first 'earthship', the Thumb House, from discarded materials and forms the architectural practice Earthship Biotecture to promote low-impact self-servicing dwellings.

—

US Congress passes Federal Water Pollution Control Amendments, later known (with additional legislation) as the Clean Water Act.

1973

Oil embargo and price rise: the first 'energy crisis'.

—

Small is Beautiful by E.F. Schumacher challenges the dominant trend towards globalisation.

—

The Ecology Party (later renamed the Green Party) founded in Britain.

—

Z for Zachariah by Robert C. O'Brien, a post-apocalyptic novel for young people.

1974

The Comedy of Survival: Joseph Meeker proposes that the truly ecological genre is comedy.

1975

Investigation of trace gases in the stratosphere leads to discovery of the danger airplane emissions pose to ozone layer.

—

In BBC TV comedy drama *The Good Life* Tom and Barbara Good give up the rat race to become self-sufficient.

—

The Monkey Wrench Gang, a novel by Edward Abbey, describes sabotage of environmentally damaging activities in the American Southwest; a possible inspiration for the founding of Earth First! in 1980.

1976

Studies show that CFCs, methane and ozone can make a serious contribution to the greenhouse effect; in 1977 the US bans CFCs from aerosol spray cans.

—

The Complete Book of Self-Sufficiency by John Seymour.

1977

Scientific opinion converges on global warming as the chief climate risk in next century.

—

Wangari Maathai founds Kenya's Green Belt Movement. By 1992 the movement has planted over seven million saplings, proving the effectiveness of grassroots organisation and 'appropriate technology' (a term popularised by Schumacher). Maathai receives the Nobel Peace Prize in 2004.

1978

The Ennead by Jan Mark,

a novel for young people in which a depleted, uninhabitable earth has been abandoned.

1979

Second oil 'energy crisis'.

—

Gaia by James Lovelock puts forward the idea that the biosphere is self-regulating.

1980

The World Conservation Strategy is published, becoming the basis for national conservation plans in many developing nations.

—

Radical direct action group Earth First! is formed by Arizona desert activists Dave Foreman, Howie Wolke and Mike Roselle.

1981

The Revenge of Samuel Stokes by Penelope Lively: ghosts of a historical garden haunt a new housing estate, culminating in the return of an ornamental lake.

1985

The UN Environment Programme, World Meteorological Organization and the International Council of Scientific Unions conference at Villach, Austria. Scientists from 29 developed and developing countries assess the role of increased carbon dioxide and other greenhouse gases and aerosols on climate changes and associated impacts. The conference is also significant in proposing that the state of scientific knowledge justified political action. Its joint statement opens: 'As a result of the increasing concentrations of greenhouse gases, it is now believed that in the first half of the next century a rise of global mean temperature could occur which is greater than any in man's history'.

—

An 'ozone hole' above Antarctica is discovered by British scientists, and is explained in terms of release of CFCs by industry and consumer products.

1986

Slow Food movement founded by Carlo Petrini in Italy.

—

In the USSR, nuclear reactor number 4 explodes in Chernobyl, Ukraine. Cultural responses include Christa Wolf's novel *Störfall* (Accident) and the play *Sarcophagus* by Vladimir Gubaryev, the science editor of *Pravda*.

1987

Gro Harlem Brundtland, Norwegian Prime Minister, defines sustainable development as 'development which meets the needs of the present without compromising the ability of future generations to meet their own needs'.

—

The Montreal Protocol, an international agreement to phase out ozone-depleting chemicals, is signed by the main industrial countries, demonstrating that fast and effective action on a global environmental issue is possible.

1988

Coverage of global warming leaps up the news agenda following record heat and droughts.

—

Dr James Hansen testifies to US Congress, saying that he could state 'with 99 per cent confidence' that there was a long-term warming trend.

—

Toronto Conference on the Changing Atmosphere calls for strict, specific limits on greenhouse gas emissions. This was the first major international meeting bringing governments and scientists together to discuss action on climate change. Industrialised countries pledge to voluntarily cut CO_2 emissions by 20% by the year 2005. This meeting is also critical in the establishment of the Intergovernmental Panel on Climate Change.

—

UK Prime Minister Margaret Thatcher is first world leader to call for action on global warming.

—

Ice-core and biology studies confirm living ecosystems give climate feedback by way of methane, which could accelerate global warming.

—

Intergovernmental Panel on Climate Change (IPCC) established to review and report on international science, impacts and responses to climate change.

—

The assassination by ranchers of Chico Mendes, leader of Brazil's rubber tappers' union, and a prominent figure in the movement to save the rainforest from illegal logging. Mendes' death was one of an estimated 1,700 resulting from land disputes in Brazil over two decades.

1989

The End of Nature by Bill McKibben.

—

Breakthrough voting levels of 15% for the UK Green Party in the European Parliament elections.

1990

IPCC publishes first Assessment Report. It finds that the planet warmed by 0.5° C in the 20th century, and warns that only strong measures to halt rising greenhouse gas emissions will prevent serious global warming.

1991

Researchers at the Centre for the Study of Environmental Change, Lancaster, look at environmental issues as sociological, cultural, philosophical phenomena, rather than as purely 'physical' or 'policy' set of issues and problems.

—

Ken Saro-Wiwa founds Nigeria's Movement for the Survival of the Ogoni People in reaction to Shell's oil drilling and extensive pollution in the Niger river delta. The country's military dictators respond to massive demonstrations with threats, intimidation and arrest of the movement's leaders and, in 1995, their execution.

1992

United Nations Conference on Environment and Development in Rio de Janeiro produces UN Framework Convention on Climate Change, but US blocks calls for serious action. Also known as Earth Summit, Rio Summit or Rio Conference.

1993

Greenland ice cores suggest that great climate changes (at least on a regional scale) can occur in the space of a single decade.

—

In Tony Kushner's play *Angels in America: Millennium Approaches*, an angel descends to earth through a hole in the ozone layer.

—

In the novel *Gridlock*, by Ben Elton, a city chokes on carbon monoxide.

—

The Poet Laureate, Ted Hughes, edits *Sacred Earth Dramas*, an anthology of plays inspired by the Earth Summit.

1995

Second IPCC report detects 'signature' of human-caused greenhouse effect warming, and declares that serious warming is likely in the coming century.

—

Reports of the breaking up of Antarctic ice shelves and other signs of warming in polar regions.

1997

International conference produces Kyoto Protocol, setting targets for industrialized nations to reduce greenhouse gas emissions if enough nations sign a treaty (rejected by the US Senate in advance).

—

Cities for a Small Planet, by architect Richard Rogers.

2000

Launch of the Ashden Directory, an online magazine about environmentalism and performing arts, with a database of productions since 1893 that have environmental themes.

—

The conference 'Between Nature: Explorations in ecology and performance' at Lancaster University brings together performers, academics and activists.

—

The Song of the Earth, a study of literature and environment by Jonathan Bate.

—

The Theft of Sita, a shadow theatre performance presenting the contemporary destruction of Indonesian rainforest through the ancient Ramayana epic.

2001

Third IPCC report states that global warming unprecedented since end of last ice age is 'very likely'.

2003

A summer heatwave kills 30,000 people in Europe. Media reports associate the highest temperatures recorded in 500 years with climate change.

—

The first Cape Farewell expedition of artists and scientists sails to the high Arctic to study global warming.

—

The Long Summer: How Climate Changed Civilisation by Brian Fagan.

—

Oryx and Crake by Margaret Atwood, a post-apocalyptic novel exploring the consequences of genetic engineering.

2004

The Day After Tomorrow, a Hollywood blockbuster that depicts a rapid transition to a new ice age.

—

The Death of Environmentalism: Global Warming Politics in a Post-Environmental World, by Michael Shellenberger and Ted Nordhaus, rejects the vision of the 1970s environmental movement and calls for radical rethinking of its aims.

—

The Noah's Ark Project details the impact of climate change on cultural heritage sites in Europe.

2005

Avoiding Dangerous Climate Change Symposium convened by the UK government to bring together the latest research on how to achieve the objective of the 1992 United Nations Framework Convention on Climate Change.

—

Kyoto treaty goes into effect, signed by major industrial nations except US.

—

Hurricane Katrina, leading to the destruction of New Orleans, and other major tropical storms in the same time period, spur debate over impact of global warming on storm intensity.

—

The Thunder Mutters: 101 Poems for the Planet, a collection of poems edited by Alice Oswald.

—

Collapse by Jared Diamond presents climate change as one of the five factors that drive social collapse.

—

First TippingPoint meeting of artists and scientists.

—

Lines of Defence, site-specific art installation by Bettina Furneé addressing coastal erosion, Bawdsey, England.

—

English Heritage report (with UCL Centre for Sustainable Heritage): *Climate Change and the Historic Environment*.

—

Northsoutheastwest exhibition of Magnum photographers, organised by the Climate Group.

—

Royal Society of the Arts 'Arts and Ecology' website launched (actively maintained until 2010).

2006

An Inconvenient Truth, a film presented by Al Gore.

—

The Stern Review on the Economics of Climate Change. Its main conclusion is that the benefits of strong, early action on climate change considerably outweigh the costs.

—

We Turned on the Light by Orlando Gough and his choral group, The Shout, commissioned for the Proms with libretto about climate change by Caryl Churchill.

—

Syriana, a political thriller about corruption in the global oil industry.

—

The Canary Project founded in the United States.

—

Transition Town Totnes founded.

—

In *Field Notes from a Catastrophe* Elizabeth Kolbert describes the impact of climate change on people in the Netherlands, Iceland and Alaska.

—

And While London Burns, a 'soundtrack for the era of climate change', an operatic audio tour by PLATFORM of institutions in London's financial district.

—

Climate Change: Cultural Change symposium by Helix Arts, at World Summit on Arts and Culture.

—

Land, Art anthology, edited by Max Andrews (RSA Arts and Ecology)

—

The Road, by Cormac McCarthy, a post-apocalyptic novel about a father and son, is widely seen as a view of the future after extreme climate change.

—

UNESCO World Heritage Centre report: *Predicting and Managing the Effects of Climate Change on World Heritage*.

—

National Trust (UK) report: *Forecast – Changeable? Climate Change Impacts around the National Trust*.

—

The Ship: The Art of Climate Change, the Cape Farewell exhibition at the Natural History Museum, London, and touring nationally and internationally thereafter.

2007

The fourth IPCC report confirms the human cause of global warming, warns that serious effects of warming have become evident, and outlines the economic and lifestyle changes necessary to mitigate those impacts. The reports state that the cost of reducing emissions would be far less than the damage they will cause.

—

Greenland and Antarctic ice sheets and Arctic Ocean sea-ice cover found to be shrinking faster than expected.

—

Floods in June force thousands in England from their homes.

—

Monsoon flooding in the Indian subcontinent causes 14 million Indians and 5 million Bangladeshis to leave their homes.

—

2,000 protestors camp on the site of a proposed third runway at Heathrow airport.

—

Arcola Energy aims to make the Arcola Theatre in London carbon neutral.

—

The National Theatre in London commits itself to reducing its use of gas and electricity by 20% over three years.

—

The Eleventh Hour, a film starring Leonardo DiCaprio.

—

Invisible Bonfires, a performance by Forkbeard Fantasy.

—

The Low Carbon Show, a regular radio programme on Resonance FM.

—

Live Earth global music event sees Al Gore and colleagues promoting action on climate change.

—

The Most Terrifying Video You'll Ever See 2, made by a US science teacher, puts a risk-based argument in a light tone, and is downloaded more than 8 million times.

—
The United States Supreme Court rules that greenhouse gases are pollutants, opening the door to litigation against industries producing high levels of carbon emissions.
—
The Carhullan Army, by Sarah Hall, a novel about a post-apocalyptic Britain whose ravages are largely due to environmental breakdown.
—
The Wild Places, by Robert Macfarlane.
—
A Moral Climate: The Ethics of Global Warming, by Michael Northcott, a philosophical/theological account of climate impacts.
—
UNESCO report: *Case Studies on Climate Change and World Heritage*.
—
The Cultures of Climate Change research group established at CRASSH, Cambridge University.
—
Weather Report: Art and Climate Change catalogue of the exhibition curated by Lucy Lippard in collaboration with EcoArts at the Boulder Museum of Contemporary Art in Colorado.
—
Ackroyd & Harvey realise *Flytower* on the exterior of London's National Theatre, growing grass on the north and west faces of the theatre's flytower.

2008

Voters in Ecuador approve a referendum on a new progressive constitution, which gives Nature the same rights as human beings.
—
Serious Things, by Gregory Norminton, a novel featuring a reclusive climate scientist on the north coast of Scotland.
—
Bipolar, collection of essays on polar regions edited by Kathryn Yusoff.
—
Future Ethics symposium, University of Manchester.
—
The London Mayor's Green Theatre - Taking Action on Climate Change programme launched, aiming to reduce by 60% the energy used by London theatres by 2025.
—
Burn Up, a thriller about the oil industry and climate change written by Simon Beaufoy, broadcast on BBC2.
—
Six Greenpeace climate change activists cleared of causing £30,000 of criminal damage at Kingsnorth coal-fired power station.
—
Asia-Europe Dialogue on Arts, Culture & Climate Change in Beijing, China gathered 43 Asian and European artists, designers, architects, cultural practitioners, environmentalists and scientists, who participated in a three-day workshop.
—
Don DeLillo, Sarah Ruhl, José Rivera, Lisa Kron and Jon Robin Baitz are among the writers of nine short plays on the theme of global warming performed at Climate of Concern, the second annual New York University Humanities Festival organised by Lawrence Weschler.

2009

'Climategate': the web publication of hacked emails written by climate scientists fuels scepticism.
—
Level of CO_2 in the atmosphere reaches 385 ppm.
—
Mean global temperature (five-year average) is 14.5 °C, the warmest in hundreds, and probably thousands of years.
—
One billion people take part in Earth Hour by switching off their lights at 8.30pm to mark the beginning of UN Climate Panel's meetings.
—
Major exhibitions on climate change and the arts: *C Words*, PLATFORM (Arnolfini, Bristol); *RETHINK* (Copenhagen); *eARTh: Art of a Changing World* (Royal Academy of Art, London); *Radical Nature* (Barbican, London); *FutureSonic* (Manchester); *Climate for Change*, (Liverpool); *Two Degrees* (ArtsAdmin, London).
—
Earth Matters on Stage, a symposium on theatre and ecology, and the Ecodrama Playwrights Festival are hosted by the University of Oregon.
—
The Contingency Plan by Steve Waters, the first play on climate change at a high-profile London theatre (Bush Theatre, London).
—
The United Nations Climate Change Conference, COP15, is held in Copenhagen, with no binding agreement reached.
—
Avatar: a film set in the mid-22nd century: mining on the planet of Pandora threatens the continued existence of a local tribe.
—
The Age of Stupid: a film directed by Franny Armstrong.
—
The LightSwitch Project, a play (also broadcast in 2011) commissioned by TippingPoint.
—
No Condition is Permanent: 19 Poets on Climate Justice and Change (PLATFORM).
—
Launch of 10:10, an organisation encouraging people, schools, businesses and organisations to cut their carbon consumption by 10% each year.

2010

Solar by Ian McEwan, a satirical novel focusing on climate change.

—

Long Horizons: An Exploration of Art and Climate Change, a collection of essays published by the British Council and Julie's Bicycle.

—

In-Flight Entertainment by Helen Simpson, a collection of climate change short stories.

—

Uncivilisation, the first Dark Mountain Festival.

—

Third Ring Out: Rehearsing the Future, an immersive performance by Metis Arts.

—

Earthquakes in London, a play by Mike Bartlett (National Theatre, London).

—

Floods in the Indus river basin of Pakistan cover one-fifth of the country's total land area and affect 20 million people.

—

Trashcatchers' Carnival, a TippingPoint commission, parades through Tooting in London.

—

The Jellyfish Theatre, a temporary structure built of reclaimed materials, opens in London with two plays on ecological themes: *Oikos* and *Protozoa*.

—

WWF publishes *Common Cause: The Case for Working with our Cultural Values*, showing the importance of values and goals in motivating change.

—

The US National Science Foundation awards a $700,000 grant to The Civilians, a New York theatre company, to finance *The Great Immensity*, a production about climate change.

—

Freedom by Jonathan Franzen, a 'green American novel', touches on mountaintop removal mining, species decline, habitat conservation, population control and suburban sprawl.

—

Commute of the Species uses puppets to chronicle the arrival of invasive animal and plant species in the Hudson River Valley while travelling on a Harlem line train from Grand Central station, New York.

—

The Department for Culture, Media and Sport in the UK publishes a Climate Change Plan.

—

The Cape Farewell exhibition *Unfold*, presenting the work of 25 artists who participated in expeditions to the High Arctic and to the Andes, opens at Kings Place, London, before touring nationally and internationally.

—

SHIFT; eight-day music-led festival, presented by Cape Farewell and Southbank Centre, London.

2011

Greenland, a play by Moira Buffini, Matt Charman, Penelope Skinner and Jack Thorne (National Theatre, London).

—

The Heretic, a play by Richard Bean (Royal Court Theatre, London).

—

Vendage Tardive, poems by Peter Reading.

—

100 Places to Remember Before they Disappear: photographs from 100 places around the world at risk of disappearing or seriously threatened by climate change.

—

The Ground Aslant: An Anthology of Radical Landscape Poetry edited by Harriet Tarlo.

—

Cave of Forgotten Dreams, a film about the Lascaux caves, directed by Werner Herzog.

—

High Arctic exhibition, National Maritime Museum, London.

The timeline was compiled in June 2011 by the editors of this volume, drawing on *The Discovery of Climate Change* by Spencer Weart (www.aip.org) and the timeline on the Ashden Directory of Environment and Performance (www.ashdendirectory.org.uk). It is part of a project to chart the history of cultural responses to climate change. The printed version is necessarily selective, and focuses mainly on Britain and the United States. To add items to the timeline, and to follow its development, please see the online version at www.ashdenizen.blogspot.com.